MCO LEVEL THREE – TEST BOOKLET

AF816476

NATIONAL
CENTER FOR
CONSTRUCTION
EDUCATION AND
RESEARCH

STORE THIS TEST BOOKLET IN A SECURE AREA!

To enhance the security and flexibility of standardized craft training testing, the National Center for Construction Education and Research offers two testing formats for the written examination. Instructors may provide testing for their trainees in one or both of the following ways:

- *By using the reproducible masters included in this booklet;*
- *By using the TestGen software to reformat existing exams and/or create final exams.*

Using the Reproducible Masters

1. Masters for the module examinations are found on the following pages. The tests can be photocopied for your local use. ***The answer keys are included in this test booklet. Be sure to keep answer keys and tests in a secure area!***
2. You may also choose the option of administering tests on overhead transparencies and having trainees fill in an answer sheet. This method improves test security since no copies of the tests are distributed.

Using the TestGen Software

The TestGen Software for *Mobile Crane Operations Level Three* is available from Prentice Hall. Detailed instructions for installing and using the software are included with the software program. The software includes all of the examination questions found in this test booklet and also allows instructors to:

- Scramble test questions.
- Add test questions.*
- Merge module examinations to create a final exam.

To order the *Mobile Crane Operations Level Three* TestGen software for only $30.00, call Prentice Hall at (800) 922-0579 and ask for ISBN# 0-13-168346-2.

Important Note:

To remain in accordance with the NCCER's accreditation policies, instructors must test trainees with NCCER's module examination. Instructors may add questions to or supplement the module exam, but ***they may not delete existing test questions****. Doing so represents a departure from NCCER Accreditation Policies and Guidelines and may seriously jeopardize either the accreditation status of the training program sponsor and/or any recognition of trainees, instructors, and trainers through the NCCER National Registry.*

NOTE ON PERFORMANCE TESTING

Performance Profile Sheet(s) are included in this Test Booklet in a format that can be easily photocopied for each trainee. This examination is designed to measure competency in the tasks taught in each module.

Please note the number of tasks to be tested while teaching each module. Each trainee should be tested on all the tasks listed on the Performance Profile Sheet(s). Before the performance testing, the instructor should brief the trainees on:

- Test objectives and criteria
- Safety precautions
- Procedures for each task to be tested

The instructor administering the performance testing should also do the following:

- Ensure that all of the needed equipment is available and operating properly.
- Set up the testing stations.
- Organize and administer the test in a way that allows for optimal performance.
- Complete the Performance Profile Sheet(s) for each trainee by assigning a pass/fail score for each listed task. Also, include the testing date for each task in the rating box.
- Monitor adherence to all safety regulations and precautions.
- Provide adequate supervision to prevent injuries.
- Take immediate and effective action to remedy any emergency.

Accreditation Testing

If Performance Testing is done as part of the National Center for Construction Education and Research Standardized Craft Training Program, the following conditions must be met:

1. The Craft Instructor must hold valid NCCER instructor certification.
2. The training must be delivered through a Training Program Sponsor recognized by the NCCER.
3. For every module, the specific performance testing must be completed to the satisfaction of the instructor.
4. The results of the testing must be recorded on the Craft Training Report Form 200. This form must be provided to the local Training Program Sponsor to be forwarded to the NCCER National Registry.

Mobile Crane Operations Level Three Test Booklet

Contains Module Examinations / Answer Keys and Performance Profile Sheets for the following Modules:

Note: Modules 21301-05 and 21304-05 contain Performance Task worksheets. There are no Performance Tasks for Module 21307-05.

Please Address Your Feedback To:

NCCER, Craft Training Department • P.O. Box 141104 • Gainesville, FL 32614-1104

Fax: 352-334-0932 Email: curriculum@nccer.org Voice: 352-334-0911

Load Charts

21301-05

Trainee Name: ____________________

Social Security Number: ____________________ **Date:** __________

_____ 1. The _____ of an object is the point in the object where its weight can be assumed to be concentrated.

a. fulcrum point
b. rotational equilibrium
c. center of gravity
d. leverage

_____ 2. The _____ is the fulcrum point on which a lever is balanced.

a. tipping axis
b. axis of rotation
c. line of balance
d. equilibrium point

_____ 3. All of the following move the crane's center of gravity toward the tipping axis *except* _____.

a. removing upperworks counterweight
b. booming down
c. telescoping out
d. adding upperworks counterweight

_____ 4. The _____ is measured from the boom foot pins to the center of the boom point sheaves.

a. load radius
b. boom length
c. boom angle
d. boom point elevation

_____ 5. In crane terminology, net capacity is the _____.

a. stowed or effective weight of the boom attachments
b. weight of the load and all attachments
c. weight of the headache ball and load block
d. gross capacity minus deductions

_____ 6. Load/capacity charts indicate the _____ lifting capacity of the crane.

a. minimum
b. mean (average)
c. net
d. gross

_____ 7. On a hydraulic boom crane, the _____ is measured between the bottom of the base section and the horizontal with the boom loaded.

a. boom angle
b. boom length
c. boom point elevation
d. load radius

_____ 8. Boom capacity, boom angle, and load operating radius values are _____.

a. determined for the crane standing level
b. determined in windy conditions to account for adverse weather
c. always in a direct relationship with each other
d. organized in the same format on every load/capacity chart

_____ 9. Capacities for crawler cranes that have outrigger extensions do *not* exceed _____ percent of tipping loads.

a. 75
b. 80
c. 85
d. 90

_____ 10. If outriggers are set but *not* extended, capacities for lifting do *not* exceed _____ percent of tipping loads.

a. 75
b. 80
c. 85
d. 90

_____ 11. Capacities for on-rubber lifting do *not* exceed _____ percent of the load needed to cause tipping.

a. 75
b. 80
c. 85
d. 90

_____ 12. To read a lattice boom crawler crane load/capacity chart, the operator must first know the _____.

a. boom length and load radius
b. boom point elevation and gross capacity
c. LMI operating code and tipping load
d. overhauling weight and line parts

_____ 13. A _____ is used as an alternative to a freestanding tower crane where a lower erection height allows its use.

a. ring attachment
b. jib
c. tower attachment
d. tower stabilizer

_____ 14. When a jib and/or boom extension is stowed on the heel section of the main boom, the stowed weights must be _____ the gross capacity in order to calculate the crane's net capacity.

a. added to
b. subtracted from
c. multiplied by
d. divided by

_____ 15. Counterweight sections for lifting are combined _____.

a. in the heaviest configuration
b. and always installed at the rear of any model crane
c. in the way specified in the load/capacity charts
d. in the configuration that allows for greater mobility around the job site

_____ 16. Stipulations regarding crane operation and capacity are contained in the _____.

a. operational notes
b. load/capacity charts
c. range diagram
d. manufacturer's web site

_____ 17. How many parts of line would be required in the following situation? Refer to *Table 1*.

- Size of wire rope: 1 inch
- Load weight: 70 tons
- Type of sheaves: anti-friction bearing

a. 4
b. 6
c. 10
d. 20

Ratio A Bronze Bushed Sheaves	Ratio B Anti-friction Bearing Sheaves	Number of Line Parts
0.96	0.98	1
1.87	1.94	2
2.75	2.88	3
3.59	3.81	4
4.39	4.71	5
5.16	5.60	6
5.90	6.47	7
6.60	7.32	8
7.27	8.16	9
7.91	8.98	10
8.52	9.79	11
9.11	10.60	12
9.68	11.40	13
10.20	12.10	14
10.70	12.90	15
11.20	13.60	16
11.70	14.30	17
12.20	15.00	18
12.60	15.70	19
13.00	16.40	20

301E01.EPS

Table 1

_____ 18. According to *ASME B30.5, Mobile and Locomotive Cranes*, manufacturers are required to provide _____.

a. load capacity information in both kilograms and pounds
b. boom angle information in a column to the left of the load radius column
c. load capacity information in a standard format
d. specific information in load rating charts for each crane

_____ 19. A side view of a crane with its full range of boom lengths, jibs, and boom extensions is a _____ diagram.

a. configuration
b. load/capacity
c. side
d. range

_____ 20. The boom angle is the angle _____.

a. between the ground and the center of the boom point sheaves
b. between the projection of the axis of rotation and the center of the vertical hoist line
c. above or below horizontal of the longitudinal axis of the base boom section
d. above or below vertical of the longitudinal axis of the base boom section

_____ 21. Telescoping out or booming down increases the load leverage by _____.

a. reducing the structural strength of the crane
b. moving the load's center of gravity closer to the tipping axis
c. moving the load's center of gravity farther from the tipping axis
d. decreasing the lifting radius

_____ 22. The longest boom lengths are found on _____ cranes.

a. industrial hydraulic
b. all-terrain
c. truck-mounted telescopic
d. lattice boom

_____ 23. When a jib is mounted on the boom tip, the effective weight must be deducted from the gross capacity _____.

a. only when the hoisting is done from the main boom
b. at all times
c. only when lifting is done from the erected jib
d. only when the boom extension is stowed on the heel section of the main boom

_____ 24. For some crane load charts, the greater the number of layers on the winch drum, the _____.

a. lower the line pull available to lift the load
b. greater the line pull available to lift the load
c. greater the overhauling weight
d. lower the overhauling weight

_____ 25. To determine the main boom angle of a telescopic boom when the boom is only partially extended and only the load radius is known, the operator refers to the _____.

a. capacity deduction chart
b. range diagram
c. boom extension diagram
d. operational notes

_____ 26. On some manufacturers' load/capacity charts, the information set off in parentheses under the gross capacities is the _____.

a. structural stability
b. boom length
c. load radius
d. boom angle

_____ 27. To make a lift, a crane's leverage has to be greater than the _____.

a. vertical distance from the center of gravity times the load leverage
b. crane weight × the horizontal distance from the center of gravity to tipping axis
c. gross capacity of the crane
d. load's leverage

_____ 28. The safe working load of a hoist rope is the diameter of the rope _____.

a. multiplied by 8
b. multiplied by 12
c. squared, then multiplied by 8
d. squared, then multiplied by 12

_____ 29. When the main boom is only partially extended and only the load radius is known, the operator can refer to the _____ to determine the unloaded boom angle of a telescopic boom.

a. range diagram
b. jib chart
c. counterweight illustration
d. reeving diagram

_____ 30. A _____ greatly increases the crane's capacity through configuration changes and weight transfer.

a. hook block
b. ring attachment
c. tower attachment
d. power-pinned fly

notes

Craft: Mobile Crane Operations

Module Number: 21301-05

Module Title: Load Charts

contren Learning Series

TRAINEE NAME: ____________________

TRAINEE SOCIAL SECURITY NUMBER: ____________________

CLASS: ____________________

TRAINING PROGRAM SPONSOR: ____________________

____

INSTRUCTOR: ____________________

Rating Levels: (1) Passed: performed task (2) Failed: did not perform task
Also, list the date the testing for each task was completed.

Recognition: When testing for the NCCER Standardized Craft Training Program, be sure to record Performance testing results on Craft Training Report Form 200, and submit the results to the Training Program Sponsor.

Objective	TASK	RATING
3	1. Identify boom angle, boom length, and load radius on a load/capacity chart.	
6	2. Identify the requirements of the on-rubber load/capacity chart.	
6	3. Identify the requirements of the on-outrigger load/capacity chart.	

continued

Craft: Mobile Crane Operations

Module Number: 21301-05

Module Title: Load Charts

contren Learning Series

Objective	TASK	RATING
6, 7, 8	4. Properly identify load/capacity charts that are used in different configurations.	
5, 8	5. Identify parts of line and counterweight considerations in load/capacity chart information.	
5	6. Calculate parts of line.	
4	7. Calculate crane capacities using load/capacity charts.	

Trainee Name: ______________________________

Social Security Number: ______________________ **Date:** ______________

Task 6 – Calculate parts of line.

_________ 1. Refer to *Table 1.* Calculate parts of line given the following:

- Size of wire rope: 1 1/4 inches
- Load weight: 45 tons
- Type of sheaves: bronze bushed

Ratio A Bronze Bushed Sheaves	Ratio B Anti-friction Bearing Sheaves	Number of Line Parts
0.96	0.98	1
1.87	1.94	2
2.75	2.88	3
3.59	3.81	4
4.39	4.71	5
5.16	5.60	6
5.90	6.47	7
6.60	7.32	8
7.27	8.16	9
7.91	8.98	10
8.52	9.79	11
9.11	10.60	12
9.68	11.40	13
10.20	12.10	14
10.70	12.90	15
11.20	13.60	16
11.70	14.30	17
12.20	15.00	18
12.60	15.70	19
13.00	16.40	20

301E01.EPS

Table 1

notes

Trainee Name: ____________________

Social Security Number: ____________________ **Date:** ____________

Task 7 – Calculate crane capacities using load/capacity charts.

For Questions 1 through 3, refer to *Figures 2* and *3*.

_________ 1. When lifting from the main boom, what is the net capacity of the Grove TM1500 in the following configuration?

- Outriggers properly set and level – over-the-rear
- Power pinned fly retracted
- Boom length of 94 ft
- Load radius of 30 ft
- 150-ton, 8-sheave hook block reeved with required parts of line
- Rigging and accessories of 180 lbs

_________ 2. When lifting from the main boom, what is the net capacity of the Grove TM1500 in the following configuration?

- Outriggers properly set and level – over-the-rear
- Power pinned fly retracted
- 33 ft–58 ft extension (retracted)
- Boom length of 106 ft
- Load radius of 35 ft
- 150-ton, 8-sheave hook block reeved with required parts of line
- Rigging and accesssories of 200 lbs

_________ 3. When lifting from the main boom, what is the net capacity of the Grove TM 1500 in the following configuration?

- Outriggers properly set and level – over-the rear
- Power pinned fly retracted
- 33 ft–58 ft extension erected (extended)
- Boom length of 88 ft
- Load radius of 20 ft
- 150-ton, 8-sheave hook block reeved with required parts of line
- Rigging and accessories of 160 lbs

Rated Lifting Capacities in Pounds
46 Ft – 173 Ft Boom on Outriggers - Over Rear

Radius in Feet	#0001									#0002
	Main Boom Length in Feet (Power Pinned Fly Retracted)									Power Pin. Fly Ext. & 141 ft.
	46	58	70	82	94	106	118	130	141	173
10	300,000 (74.5)									
12	280,000 (72)	143,500 (76)	142,000 (79)							
15	235,000 (67.5)	143,500 (72.5)	141,500 (76.5)	130,000 (78.5)						
20	173,500 (60.5)	143,500 (67.5)	123,500 (72)	112,000 (75)	102,000 (77.5)	90,300 (79.5)				
25	136,500 (52)	131,500 (61.5)	110,500 (67.5)	98,650 (71)	89,250 (74)	78,550 (76.5)	73,700 (78.5)	69,300 (80)		
30	106,000 (43)	106,000 (55.5)	98,000 (63)	88,350 (67.5)	78,750 (71)	69,250 (73.5)	65,100 (76)	61,000 (77.5)	60,000 (79.5)	
35	84,700 (30.5)	84,700 (49)	84,700 (58)	80,150 (63.5)	69,000 (67.5)	60,750 (70.5)	57,150 (73)	54,000 (75.5)	52,150 (77.5)	
40		70,500 (41)	70,500 (52.5)	70,500 (59.5)	61,300 (64)	54,000 (67.5)	50,600 (70.5)	48,300 (73)	45,850 (75)	38,000 (79)
45		58,850 (32)	58,850 (47)	58,850 (55)	55,000 (60.5)	48,500 (64.5)	46,200 (68)	43,050 (71)	40,400 (73)	35,750 (77)
50		49,600 (17.5)	49,600 (40.5)	49,600 (50.5)	48,750 (57)	43,050 (61.5)	40,700 (65)	38,250 (68.5)	35,750 (71)	32,100 (75.5)
60			36,200 (22.5)	36,200 (39.5)	36,200 (48.5)	34,300 (55)	33,600 (59.5)	30,750 (63.5)	28,500 (66.5)	26,350 (72)
70				26,050 (25)	26,050 (39.5)	26,050 (47.5)	26,050 (53)	24,750 (58)	23,100 (61.5)	22,000 (68.5)
80					18,850 (27)	18,850 (39)	18,850 (46.5)	18,850 (52.5)	18,700 (56.5)	18,500 (64.5)
90						13,500 (28)	13,500 (38.5)	13,500 (46.5)	13,500 (51.5)	15,250 (60.5)
100							9,390 (29)	9,390 (38)	9,390 (45.5)	12,600 (56.5)
110							6,080 (12.5)	6,080 (30.5)	6,080 (39)	10,100 (52)
120								3,390 (17.5)	3,390 (31)	7,530 (47.5)
130									1,150 (19.5)	5,390 (42.5)
140										3,610 (36.5)
150										2,100 (30)
Minimum boom angle (deg.) for indicated length (no load)									10	19
Maximum boom length (ft.) at 0 deg. boom angle (no load)									140	167

Note: ( ) Boom angles are in degrees.
#LMI operating code. Refer to LMI manual for instructions.

301E02.EPS

Figure 1

WEIGHT REDUCTIONS FOR LOAD HANDLING DEVICES

33 ft. Extension	
*Stowed -	892 lbs.
*Erected -	5,704 lbs.

33 ft. - 58 ft. Extension	
*Stowed -	1,246 lbs.
*Erected (Ret.) -	8,412 lbs.
*Erected (Ext.) -	11,406 lbs.

46 ft. - 173 ft. Boom with	
*46 ft. Jib Erected -	9,613 lbs.
*60 ft. Jib Erected -	14,571 lbs.
*74 ft. Jib Erected -	20,443 lbs.
*88 ft. Jib Erected -	27,199 lbs.
*Fixed Jib Accessories -	327 lbs.

*Reduction of main boom capacities

HOOKBLOCKS:	
30 Ton, 1 Sheave	1,022 lbs.
150 Ton, 8 Sheave	5,254 lbs.
Auxiliary Boom Head	261 lbs.
10 Ton Headache Ball	560 lbs.
15 Ton Headache Ball	803 lbs.

301E03.EPS

Figure 2

For Questions 4 through 9, refer to *Figures 3* through *6*.

__________ 4. What is the net capacity of the Manitowoc 4100W-2 in the following configuration?

- 22C boom, 180-ft. main boom no offset
- 115 ft of load radius
- Single-sheave upper boom point attached
- 15-ton hook block, 15 ft below boom point
- 100-ton block reeved with 4 parts of line
- Crawlers extended
- Rigging of 345 lbs
- Lifting from the main hoist

__________ 5. What is the net capacity of the Manitowoc 4100W-2 in the following configuration?

- 22C boom, 220-ft main boom no offset
- 155 ft of load radius
- 2-sheave boom point attached
- 15-ton hook block, 30 ft below boom point
- 100-ton block reeved with 2 parts of line
- Crawlers extended
- Rigging of 230 lbs
- Lifting from the main hoist

__________ 6. What is the net capacity of the 4100W-2 in the following configuration?

- 22C boom, 210 ft main boom no offset
- 135 ft load radius
- 2-sheave boom point attached
- 15-ton hook block 20 ft below boom point
- 100-ton block reeved with 4 parts of line
- Crawlers extended
- Rigging of 245 lbs

__________ 7. What is the maximum radius at which a net load of 168,000 lbs. can be safely handled by the 4100W-2 when it is configured as follows?

- 22C boom with open-throat top 160 ft in length
- 100-ton block from main boom reeved to 6 parts of line
- Rigging of 875 lbs

__________ 8. What is the gross capacity of the Manitowoc 4100W-2 in the following configuration?

- 22C open-throat top boom
- 180 ft of main boom
- Boom angle 60°

_____________ 9. What is the minimum boom angle that can be used with the 4100W-2 lifting a gross load of 14,800 lbs with the machine in the following configuration?

- 22C boom with open-throat top
- 260 ft of main boom

MANITOWOC 4100 W-2

LIFTCRAN CAPACITIES

BOOM NO. 22C WITH OPEN THROAT TOP
146,400 LB. CRANE COUNTERWEIGHT
60,000 LB. CARBODY COUNTERWEIGHT
26'6" CRAWLERS EXTENDED

***WARNING:** This chart will apply only when two 12,000 lb. side ctwts. and two 30,000 lb. carbody ctwts. bear MEC registered Serial Numbers.*

LIFTING CAPACITIES: Capacities for various boom lengths and operating radii may be based on percent of tipping, strength of structural components, operating speeds and other factors.

Capacities are for freely suspended loads and do not exceed 75% of a static tipping load. Capacities based on structural competence are shown by shaded areas.

Capacities are shown in pounds. Deduct 1200 pounds from capacities listed when single sheave upper boom point is attached and 1500 pounds when two sheave upper boom point is attached. To comply with B30.5 requirements, upper boompoint cannot be used on the 260 ft. boom. Weight of jib, (see chart A), all load blocks, hooks, weight ball, slings, hoist lines beneath boom and jib point sheaves, etc., is considered part of the main boom load. Boom is not to be lowered beyond radii where combined weights are greater than rated capacity. Where no capacity is shown, operation is not intended or approved.

OPERATING CONDITIONS: Machine to operate in a level position on a firm surface with crawlers fully extended and gantry in working position and be rigged in accordance with and under conditions referred to in rigging drawing No. 190693 and load line specification chard No. 6592-A.

Crane operator judgement must be used to allow for dynamic load effects of swinging, hoisting or lowering, travel, as well as adverse operating conditions & physical machine depreciation.

OPERATOR RADIUS: Operating is the horizontal distance form the axis of rotation to the center of vertical hoist line or load block with the load freely suspended. Add 14" to boom point radius for radius of sheave when using single part hoist line.

Boom angle is the angle between horizontal and centerline of boom butt and inserts and is an indication of operating radius. In all cases, operating radius shall govern capacity.

BOOM POINT ELEVATION: Boom point elevation, in feet, is the vertical distance from ground level to centerline of boom point shaft.

MACHINE EQUIPMENT: Machine equipped with 26'6" extendible crawlers, 48" treads, 17' retractable gantry, 12 part boom hoist reeving, four 1 3/8" boom pendants, 1st ctwt. 41,900 lbs., 2nd ctwt. 41,500 lbs., 3rd ctwt. 39,000 lbs., two 12,000 lbs. side ctwt's. and two 30,000 lbs. carbody ctwt's.

LOAD AND WHIP LINE SPECIFICATIONS	
LOAD LINE:	1-1/8" - 6 x 31 Warrington-Seale, Extra Improved Plow Steel, Regular Lay, IWRC. Minimum Breaking Strength 65 Ton. (Approx.Weight Per Ft. in Lbs. 2.34)
WHIP LINE:	1-1/8" - Warrington-Seale, Improved Plow Steel, Regular Lay, IWRC. Minimum Breaking Strength 56.5 Ton. Maximum Load - 28,300 Lbs. Per Line. (Approx. Weight Per Ft. in Lbs. 2.34)

MAXIMUM BOOM AND JIB LENGTH LIFTED UNASSISTED				DEDUCT FROM CAPACITIES WHEN JIB IS ATTACHED	
OVER FRONT OF BLOCKED CRAWLERS		OVER SIDE OF EXTENDED CRAWLERS			
BOOM LENGTH	JIB NO. 123	BOOM LENGTH	JIB NO. 123	JIB LENGTH	JIB NO. 123
260'	--	260'	--	30'	3,000 lbs.
250'	--	250'	--	40'	3,600 lbs.
240'	40'	240'	40'	50'	4,200 lbs.
230'	60'	230'	60'	60'	4,900 lbs.
Load block, hook and weight ball on ground to start.					

HOIST REEVING FOR MAIN LOAD BLOCK						
No.Parts Of Line	1	2	3	4	5	6
Max Load - Lbs.	32,500	65,000	97,500	130,000	162,500	195,000
No. Parts of Line	7	8	9	10	11	12
Max. Load - Lbs.	227,500	260,000	292,500	325,000	357,500	400,000
No. Parts of Line	13					
Max. Load - Lbs.	430,000					

FOR JIB CAPACITIES, CONSULT JIB CHART.

BOOM LGTH FEET	OPER. RAD. FEET	BOOM ANG. DEG.	BOOM POINT ELEV.	CAPACITY: CRAWLERS EXTENDED
70	16.5	79.7	75.9	460,000
	17	79.3	75.8	400,000
	18	78.5	75.6	380,100
	19	77.6	75.4	363,000
	20	76.8	75.1	347,300
	22	75.1	74.6	319,600
	24	73.4	74.1	293,400
	26	71.7	73.5	266,100
	28	69.9	72.8	237,500
	30	68.2	72.0	214,300
	32	66.4	71.2	195,100
	34	64.6	70.2	178,900
	36	62.8	69.3	165,200
	38	60.9	68.2	153,300
	40	59.1	67.0	143,000
	45	54.1	63.7	122,100
	50	48.9	59.8	106,300
	55	43.2	54.9	93,900
	60	36.9	49.0	84,000
	65	29.4	41.3	75,800
	70	19.5	30.3	63,900

BOOM LGTH FEET	OPER. RAD. FEET	BOOM ANG. DEG.	BOOM POINT ELEV.	CAPACITY: CRAWLERS EXTENDED
80	17	80.6	85.9	392,800
	18	79.9	85.8	378,900
	19	79.2	85.6	361,800
	20	78.5	85.4	346,100
	22	77.0	84.9	318,400
	24	75.5	84.5	292,500
	26	74.0	83.9	265,600
	28	72.5	83.3	237,000
	30	71.0	82.7	213,800
	32	69.5	81.9	194,600
	34	68.0	81.2	178,500
	36	66.4	80.3	164,700
	38	64.8	79.4	152,800
	40	63.3	78.4	142,400
	45	59.2	75.7	121,500
	50	54.9	72.5	105,700
	55	50.4	68.6	93,300
	60	45.6	64.1	83,400
	65	40.3	58.8	75,200
	70	34.4	52.2	68,300
	75	27.4	43.9	52,500
	80	18.2	32.0	53,900

BOOM LGTH FEET	OPER. RAD. FEET	BOOM ANG. DEG.	BOOM POINT ELEV.	CAPACITY: CRAWLERS EXTENDED
90	18	81.1	95.9	355,400
	19	80.4	95.7	346,900
	20	79.8	95.6	336,900
	22	78.5	95.2	317,400
	24	77.2	94.7	291,700
	26	75.9	94.3	264,800
	28	74.5	93.7	236,600
	30	73.2	93.2	213,400
	32	71.9	92.5	194,200
	34	70.5	91.9	178,000
	36	69.2	91.1	164,200
	38	67.8	90.3	152,300
	40	66.4	89.5	142,000
	45	62.9	87.1	121,100
	50	59.3	84.4	105,200
	55	55.5	81.2	92,800
	60	51.5	77.5	82,900
	65	47.3	73.2	74,700
	70	42.8	68.2	67,800
	75	37.9	62.3	62,000
	80	32.4	55.2	57,000
	85	25.8	46.2	52,600
	90	17.1	33.5	45,900

BOOM LGTH FEET	OPER. RAD. FEET	BOOM ANG. DEG.	BOOM POINT ELEV.	CAPACITY: CRAWLERS EXTENDED
100	19	81.4	105.9	332,900
	20	80.8	105.7	327,100
	22	79.6	105.4	316,200
	24	78.5	105.0	290,800
	26	77.3	104.5	263,900
	28	76.1	104.1	236,200
	30	74.9	103.6	212,900
	32	73.7	103.0	193,700
	34	72.5	102.4	177,500
	36	71.3	101.7	163,700
	38	70.1	101.0	151,800
	40	68.9	100.3	141,400
	45	65.8	98.2	120,500
	50	62.6	95.8	104,700
	55	59.3	93.0	92,300
	60	55.9	89.8	82,300
	65	52.4	86.2	74,100
	70	48.7	82.1	67,200
	75	44.8	77.4	61,400
	80	40.5	72.0	56,400
	85	35.9	65.6	52,000
	90	30.7	58.0	48,200
	95	24.5	48.5	44,900
	100	16.3	35.0	39,300

CAUTION! CHECK AMOUNT OF COUNTERWEIGHT ON MACHINE BEFORE USE OF THIS CHART

301E04A.EPS

Figure 3

LIFTCRANE CAPACITIES – 4100W SERIES 2 BOOM NO 22C WITH OPEN-THROAT TOP, CONTINUED

BOOM LGTH FEET	OPER. RAD. FEET	BOOM ANG. DEG.	BOOM POINT ELEV.	CAPACITY: CRAWLERS EXTENDED
110	22	82.3	115.7	240,500
	24	81.3	115.4	232,100
	26	80.2	115.0	224,400
	28	79.1	114.6	217,300
	30	78.1	114.1	210,700
	32	77.0	113.6	193,300
	34	75.9	113.1	177,100
	36	74.8	112.5	163,300
	38	73.7	111.8	151,400
	40	72.6	111.2	141,000
	45	69.9	109.3	120,100
	50	67.0	107.2	104,200
	55	64.1	104.7	91,800
	60	61.2	101.9	81,800
	65	58.1	98.8	73,600
	70	54.9	95.3	66,800
	75	51.6	91.3	60,900
	80	48.1	86.8	55,900
	85	44.4	81.7	51,600
	90	40.4	75.9	47,800
	95	36.1	69.2	44,400
	100	31.1	61.1	41,400
	105	25.3	51.1	38,700
	110	17.6	37.2	33,900
120	22	83.0	125.9	228,700
	24	82.0	125.5	220,500
	26	81.0	125.2	213,000
	28	80.0	124.8	206,100
	30	79.1	124.4	199,700
	32	78.1	123.9	192,800
	34	77.1	123.4	176,600
	36	76.1	122.9	162,800
	38	75.1	122.3	150,900
	40	74.1	121.7	140,500
	45	71.6	120.0	119,500
	50	69.1	118.1	103,700
	55	66.5	115.9	91,300
	60	63.8	113.4	81,300
	65	61.1	110.6	73,100
	70	58.3	107.5	66,200
	75	55.4	104.0	60,400
	80	52.4	100.1	55,300
	85	49.2	95.8	51,000
	90	45.9	91.0	47,200
	95	42.4	85.5	43,800
	100	38.6	79.4	40,800
	105	34.4	72.2	38,100
	110	29.7	63.7	35,700
	115	24.1	53.2	33,500
130	24	82.6	135.7	212,700
	26	81.7	135.3	205,300
	28	80.8	135.0	198,600
	30	79.9	134.6	192,300
	32	79.0	134.2	186,500
	34	78.1	133.7	176,300
	36	77.2	133.2	162,500
	38	76.3	132.7	150,600
	40	75.4	132.1	140,200
	45	73.1	130.6	119,300
	50	70.8	128.8	103,400
	55	68.4	126.8	91,000
	60	66.0	124.6	81,000
	65	63.5	122.0	72,800
	70	61.0	119.2	65,900
	75	58.4	116.1	60,000
	80	55.7	112.7	55,000
	85	53.0	108.9	50,700
	90	50.1	104.8	46,800
	95	47.1	100.1	43,500
	100	43.9	95.0	40,500
	105	40.6	89.2	37,800
	110	36.9	82.6	35,300
	115	33.0	75.1	33,100
	120	28.5	66.1	31,100
	125	23.1	55.1	29,300
140	26	82.3	145.5	198,400
	28	81.5	145.1	191,800
	30	80.7	144.8	185,600
	32	79.8	144.4	179,900
	34	79.0	144.0	174,700
	36	78.1	143.5	162,000
	38	77.3	143.0	150,100
	40	76.5	142.5	139,700
	45	74.3	141.1	118,700
	50	72.2	139.4	102,800
	55	70.0	137.6	90,400
	60	67.8	135.5	80,400
	65	65.6	133.2	72,200
	70	63.3	130.7	65,300
	75	60.9	127.9	59,500
	80	58.5	124.8	54,400
	85	56.1	121.4	50,100
	90	53.5	117.7	46,200
	95	50.9	113.7	42,900
	100	48.1	109.2	39,900
140	105	45.3	104.2	37,200
	110	42.2	98.8	34,700
	115	39.0	92.7	32,500
	120	35.5	85.7	30,500
	125	31.7	77.8	28,700
	130	27.4	68.5	27,000
	135	22.2	57.0	25,500
150	28	82.1	155.3	185,600
	30	81.3	154.9	179,600
	32	80.5	154.6	174,000
	34	79.7	154.2	168,800
	36	79.0	153.8	161,600
	38	78.2	153.3	149,600
	40	77.4	152.8	139,200
	45	75.4	151.5	118,200
	50	73.4	150.0	102,400
	55	71.4	148.3	89,900
	60	69.4	146.4	79,900
	65	67.3	144.3	71,700
	70	65.2	141.9	64,800
	75	63.0	139.4	59,000
	80	60.9	136.6	53,900
	85	58.6	133.5	49,600
	90	56.3	130.1	45,700
	95	54.0	126.5	42,400
	100	51.5	122.5	39,400
	105	49.0	118.2	36,700
	110	46.4	113.4	34,200
	115	43.6	108.2	32,000
	120	40.7	102.4	30,000
	125	37.6	96.0	28,200
	130	34.2	88.8	26,500
	135	30.5	80.5	25,000
	140	26.4	70.8	23,500
	145	21.4	58.8	22,200
160	28	82.6	165.4	179,800
	30	81.8	165.1	173,900
	32	81.1	164.7	168,400
	34	80.4	164.4	163,300
	36	79.7	164.0	158,600
	38	78.9	163.6	149,100
	40	78.2	163.1	138,700
	45	76.3	161.9	117,700
	50	74.5	160.5	101,800
	55	72.6	158.9	89,300
	60	70.7	157.1	79,300
	65	68.8	155.1	71,100
	70	66.8	153.0	64,200
	75	64.9	150.6	58,400
	80	62.9	148.0	53,300
	85	60.8	145.2	49,000
	90	58.7	142.2	45,100
	95	56.6	138.8	41,800
	100	54.4	135.2	38,700
	105	52.1	131.4	36,000
	110	49.7	127.1	33,600
	115	47.3	122.5	31,400
	120	44.8	117.5	29,400
	125	42.1	112.0	27,600
	130	39.3	105.9	25,900
	135	36.3	99.2	24,300
	140	33.1	91.7	22,900
	145	29.5	83.0	21,600
	150	25.5	72.9	20,400
	155	20.7	60.5	19,200
170	30	82.3	175.2	169,000
	32	81.6	174.9	163,600
	34	81.0	174.5	158,700
	36	80.3	174.2	154,000
	38	79.6	173.8	148,800
	40	78.9	173.3	138,400
	45	77.2	172.2	117,400
	50	75.4	170.9	101,500
	55	73.7	169.4	89,100
	60	71.9	167.7	79,000
	65	70.1	165.9	70,800
	70	68.3	163.9	63,900
	75	66.4	161.7	58,100
	80	64.6	159.3	53,000
	85	62.7	156.7	48,700
	90	60.8	153.9	44,900
	95	58.8	150.8	41,500
	100	56.8	147.5	38,500
	105	54.7	144.0	35,800
	110	52.6	140.2	33,300
	115	50.4	136.0	31,100
	120	48.1	131.5	29,100
	125	45.8	126.7	27,300
	130	43.3	121.4	25,600
	135	40.8	115.7	24,100
	140	38.1	109.3	22,600
	145	35.2	102.3	21,300
	150	32.0	94.5	20,100
	155	28.6	85.5	18,900
	160	24.7	75.1	17,800
	165	20.0	62.2	16,400
180	32	82.1	185.0	158,700
	34	81.5	184.7	154,000
	36	80.8	184.3	149,500
	38	80.2	184.0	145,200
	40	79.5	183.6	137,900
	45	77.9	182.5	116,900
	50	76.3	181.2	101,000
	55	74.6	179.8	88,500
	60	72.9	178.3	78,500
	65	71.3	176.6	70,300
	70	69.6	174.7	63,400
	75	67.8	172.6	57,500
	80	66.1	170.4	52,500
	85	64.3	168.0	48,100
	90	62.5	165.4	44,300
	95	60.7	162.5	40,900
	100	58.9	159.5	37,900
	105	57.0	156.2	35,200
	110	55.0	152.7	32,700
	115	53.0	149.0	30,500
	120	51.0	144.9	28,500
	125	48.9	140.5	26,700
	130	46.7	135.8	25,000
	135	44.4	130.7	23,500
	140	42.0	125.2	22,000
	145	39.6	119.2	20,700
	150	36.9	112.6	19,500
	155	34.1	105.3	18,300
	160	31.1	97.2	17,200
	165	27.7	87.9	16,200
	170	23.9	77.1	15,300
	175	19.4	63.8	13,500
190	32	82.5	195.1	153,100
	34	81.9	194.8	148,800
	36	81.3	194.5	144,800
	38	80.7	194.1	140,800
	40	80.1	193.8	137,000
	45	78.5	192.7	116,400
	50	77.0	191.5	100,500
	55	75.4	190.2	88,000
	60	73.9	188.8	78,000
	65	72.3	187.1	69,700
	70	70.7	185.4	62,800
	75	69.1	183.4	57,000
	80	67.4	181.4	52,000
	85	65.8	179.1	47,600
	90	64.1	176.6	43,800
	95	62.4	174.0	40,400
	100	60.7	171.2	37,400
	105	58.9	168.2	34,700
	110	57.1	164.9	32,200
	115	55.3	161.4	30,000
	120	53.4	157.7	28,000
	125	51.5	153.7	26,200
	130	49.5	149.5	24,500
	135	47.5	144.9	22,900
	140	45.3	140.0	21,500
	145	43.1	134.6	20,200
	150	40.8	128.9	18,900
	155	38.4	122.6	17,800
	160	35.9	115.8	16,700
	165	33.2	108.3	15,700
	170	30.2	99.8	14,800
	175	27.0	90.3	13,900
	180	23.3	79.1	13,100
	185	18.9	65.4	11,000
200	34	82.3	204.9	143,100
	36	81.7	204.6	139,500
	38	81.2	204.3	136,000
	40	80.6	203.9	132,500
	45	79.1	203.0	115,900
	50	77.7	201.8	100,000
	55	76.2	200.6	87,500
	60	74.7	199.2	77,400
	65	73.2	197.7	69,200
	70	71.7	196.0	62,300
	75	70.2	194.2	56,400
	80	68.6	192.2	51,400
	85	67.1	190.1	47,000
	90	65.5	187.8	43,200
	95	63.9	185.3	39,800
	100	62.3	182.7	36,800
	105	60.6	179.8	34,100
	110	59.0	176.8	31,600
	115	57.3	173.6	29,400
	120	55.5	170.2	27,400
	125	53.0	166.5	25,600
	130	51.9	162.6	23,900
	135	50.1	158.4	22,300
	140	48.1	153.9	20,900
	145	46.2	149.1	19,600
	150	44.1	144.0	18,300
	155	42.0	138.4	17,200
	160	39.7	132.5	16,100
	165	37.4	126.0	15,100
	170	34.9	118.9	14,200
200	175	32.3	111.1	13,300
	180	29.4	102.4	12,400
	185	26.2	92.5	11,700
	190	22.6	81.0	10,900
	195	18.4	67.0	8,500
210	36	82.1	214.7	134,200
	38	81.6	214.4	131,000
	40	81.0	214.1	127,900
	45	79.6	213.2	115,500
	50	78.3	212.1	99,600
	55	76.9	210.9	87,100
	60	75.4	209.6	77,100
	65	74.0	208.1	68,800
	70	72.6	206.6	61,900
	75	71.2	204.8	56,100
	80	69.7	203.0	51,000
	85	68.2	201.0	46,700
	90	66.7	198.8	42,800
	95	65.2	196.5	39,400
	100	63.7	194.0	36,400
	105	62.2	191.3	33,700
	110	60.6	188.5	31,300
	115	59.0	185.5	29,100
	120	57.4	182.3	27,100
	125	55.7	178.9	25,200
	130	54.1	175.2	23,500
	135	52.3	171.4	22,000
	140	50.6	167.3	20,600
	145	48.8	162.9	19,200
	150	46.9	158.2	18,000
	155	45.0	153.2	16,800
	160	43.0	147.9	15,800
	165	40.9	142.1	14,800
	170	38.7	135.9	13,800
	175	36.5	129.2	12,900
	180	34.0	121.9	12,100
	185	31.5	113.9	11,300
	190	28.7	104.9	10,600
	195	25.6	94.7	9,900
	200	22.1	82.9	8,800
	205	17.9	68.5	6,500
220	38	82.0	224.6	125,700
	40	81.4	224.2	122,800
	45	80.1	223.3	115,000
	50	78.8	222.3	99,100
	55	77.5	221.2	86,600
	60	76.1	220.0	76,600
	65	74.8	218.6	68,300
	70	73.4	217.1	61,400
	75	72.0	215.4	55,500
	80	70.7	213.7	50,500
	85	69.3	211.8	46,100
	90	67.9	209.7	42,300
	95	66.4	207.5	38,900
	100	65.0	205.2	35,800
	105	63.6	202.7	33,100
	110	62.1	200.0	30,700
	115	60.6	197.2	28,500
	120	59.1	194.2	26,500
	125	57.5	191.0	24,600
	130	55.9	187.6	23,000
	135	54.3	184.0	21,400
	140	52.7	180.2	20,000
	145	51.0	176.1	18,600
	150	49.3	171.8	17,400
	155	47.6	167.3	16,300
	160	45.7	162.4	15,200
	165	43.9	157.2	14,200
	170	41.9	151.7	13,200
	175	39.9	145.7	12,300
	180	37.8	139.3	11,500
	185	35.6	132.4	10,700
	190	33.2	124.8	10,000
	195	30.7	116.6	9,300
	200	28.0	107.3	8,600
	205	25.0	96.9	8,000
	210	21.5	84.8	6,500

CAPACITIES CONTINUED ON NEXT PAGE

These load charts are intended for instructional purposes only. They were derived from manufacturer sales information which may not be complete or machine specific. Not responsible for typographical errors.

301E04B.EPS

Figure 4

LIFTCRANE CAPACITIES – 4100W SERIES 2 BOOM NO 22C WITH OPEN-THROAT TOP, CONTINUED

SEE CONDITIONS ON FRONT PAGE

BOOM LGTH FEET	OPER. RAD. FEET	BOOM ANG. DEG.	BOOM POINT ELEV.	CAPACITY: CRAWLERS EXTENDED
230	38	81.5	234.5	132,900
	40	81.0	234.2	125,200
	45	79.7	233.3	114,500
	50	78.5	232.4	98,600
	55	77.2	231.3	86,100
	60	75.9	230.1	76,100
	65	74.6	228.8	67,800
	70	73.3	227.3	60,900
	75	72.0	225.8	55,000
	80	70.7	224.1	50,000
	85	69.4	222.3	45,600
	90	68.0	220.3	41,700
	95	66.7	218.2	38,400
	100	65.3	216.0	35,300
	105	64.0	213.6	32,600
	110	62.6	211.1	30,200
	115	61.1	208.4	28,000
	120	59.7	205.6	26,000
	125	58.3	202.6	24,100
	130	56.8	199.4	22,400
	135	55.3	196.0	20,900
	140	53.8	192.5	19,400
	145	52.2	188.7	18,100
	150	50.6	184.7	16,900
	155	49.0	180.5	15,700
	160	47.3	176.0	14,700
	165	45.6	171.3	13,600
	170	43.8	166.2	12,700
	175	42.0	160.8	11,800
	180	40.1	155.1	11,000
	185	38.1	148.9	10,200
	190	36.0	142.3	9,400
	195	33.9	135.1	8,700
	200	31.6	127.3	8,100
	205	29.1	118.8	7,400
	210	26.4	109.3	6,800
240	40	81.4	244.3	123,400
	45	80.2	243.5	112,600
	50	78.9	242.6	98,100
	55	77.7	241.5	85,600
	60	76.5	240.4	75,600
	65	75.3	239.1	67,300
	70	74.0	237.7	60,400
	75	72.8	236.3	54,500
	80	71.5	234.6	49,500
	85	70.3	232.9	45,100
	90	69.0	231.1	41,200
	95	67.7	229.1	37,800
	100	66.4	227.0	34,800
	105	65.1	224.7	32,100
	110	63.8	222.3	29,700
	115	62.5	219.8	27,400
	120	61.1	217.1	25,400
	125	59.7	214.3	23,600
	130	58.3	211.3	21,900
	135	56.9	208.1	20,300
	140	55.5	204.7	18,900
	145	54.0	201.2	17,600
	150	52.5	197.5	16,300
	155	51.0	193.5	15,200
	160	49.5	189.4	14,100
	165	47.9	185.0	13,100
	170	46.2	180.3	12,200
	175	44.6	175.4	11,300
	180	42.8	170.2	10,400
	185	41.0	164.6	9,600
	190	39.2	158.7	8,900
	195	37.3	152.3	8,200
	200	35.2	145.5	7,500
	205	33.1	138.1	6,700
	210	30.9	130.1	6,000

BOOM LGTH FEET	OPER. RAD. FEET	BOOM ANG. DEG.	BOOM POINT ELEV.	CAPACITY: CRAWLERS EXTENDED
250	45	80.6	253.6	106,600
	50	79.4	252.7	97,800
	55	78.2	251.7	85,200
	60	77.1	250.6	75,200
	65	75.9	249.4	66,900
	70	74.7	248.1	60,000
	75	73.5	246.7	54,100
	80	72.3	245.2	49,100
	85	71.1	243.5	44,700
	90	69.9	241.7	40,800
	95	68.7	239.8	37,400
	100	67.4	237.8	34,400
	105	66.2	235.7	31,700
	110	64.9	233.4	29,300
	115	63.6	231.0	27,100
	120	62.4	228.5	25,000
	125	61.1	225.8	23,200
	130	59.7	222.9	21,500
	135	58.4	219.9	20,000
	140	57.0	216.8	18,500
	145	55.7	213.4	17,200
	150	54.3	209.9	16,000
	155	52.8	206.2	14,800
	160	51.4	202.4	13,700
	165	49.9	198.3	12,700
	170	48.4	193.9	11,800
	175	46.8	189.4	10,900
	180	45.3	184.5	10,100
	185	43.6	179.4	9,300
	190	41.9	174.0	8,500
	195	40.2	168.3	7,800
	200	38.4	162.2	7,000
	205	36.5	155.7	6,200
	210	34.5	148.6	5,500
260	45	80.9	263.7	104,800
	50	79.8	262.9	95,700
	55	78.7	261.9	84,700
	60	77.6	260.9	74,700
	65	76.4	259.7	66,400
	70	75.3	258.5	59,500
	75	74.2	257.1	53,600
	80	73.0	255.6	48,500
	85	71.8	254.1	44,100
	90	70.7	252.4	40,300
	95	69.5	250.6	36,900
	100	68.3	248.6	33,900
	105	67.1	246.6	31,100
	110	65.9	244.4	28,700
	115	64.7	242.1	26,500
	120	63.5	239.7	24,500
	125	62.3	237.1	22,600
	130	61.0	234.4	20,900
	135	59.7	231.6	19,400
	140	58.5	228.6	17,900
	145	57.2	225.4	16,600
	150	55.8	222.1	15,400
	155	54.5	218.7	14,200
	160	53.1	215.0	13,100
	165	51.7	211.2	12,100
	170	50.3	207.1	11,200
	175	48.9	202.9	10,300
	180	47.4	198.4	9,500
	185	45.9	193.7	8,700
	190	44.3	188.7	7,900
	195	42.7	183.4	7,000
	200	41.1	177.8	6,200
	205	39.4	171.9	5,400
	210	37.6	165.6	4,700

301E04C.EPS

Figure 5

WEIGHTS 4100W Cont'd

DESCRIPTION	APPROX. WEIGHT (IN LBS.)
JIB NO. 123	
Jib Top - 15' (w/Jib point)	695
Jib Butt - 15'	690
Jib Insert - 10'	340
Basic Pendant - 33' 3-3/4" (2 Req'd)	115 each
Pendant - 10' (2 Per insert)	65 each
Jib Backstay Pendant	155 each
Jib Strut - 12'- 6"	365
COMPONENTS	
Hook Rollers (6) - w/shafts	1,020
Light Plant - 6.5KW - w/mounting platform	1,390
Catwalk - left and right side w/rails	1,320
Lagging - 27-5/8" dia. plain	1,410
Boom Hoist Rope - 12 Part - 760' of 7/8" - 6 x 26	1,080
Wire Rope Guide Assembly - Lower	325
Wire Rope Guide Assembly - Upper	510
Rope Guide Roller Assembly	55 each
2-Part Gantry w/Telescopic Backhitch	7,805
Equalizer	2,000
Hoist Line - 1-1/8" - 6 x 31	2.34 lbs./ft.
Whip Line - 1-1/8" - 6 x 31	2.34 lbs./ft.
15-Ton Hook and Weight Ball	865
100-Ton Hook Block Assembly	2,065
200-Ton Hook Block Assembly	4,900
230-Ton Hook Block Assembly	5,375
Boom Stop - Telescopic Air Cushioned	675
Dragline Fairlead - Revolving	1,910
Dragline Fairlead - Hinged	9,330

NOTE: The above weights may fluctuate up or down 5% due to manufacturing tolerances.

301E04D.EPS

Figure 6

notes

ANSWER KEY

Answer	Section Reference
1. c	2.0.0
2. a	2.0.0; Glossary
3. d	2.0.0
4. b	4.5.0; Glossary
5. d	5.0.0; Glossary
6. d	5.0.0
7. a	5.0.0; Glossary
8. a	5.0.0
9. c	6.3.3
10. a	6.3.2
11. a	6.3.1
12. a	6.1.0
13. c	6.4.0
14. b	6.6.1
15. c	6.5.0
16. a	7.0.0
17. c*	6.6.2
18. d	6.1.0
19. d	8.0.0
20. c	5.0.0; Glossary
21. c	2.0.0
22. d	4.5.0
23. a	6.6.1
24. a	6.6.2
25. b	8.0.0
26. d	6.1.0; Figures 17 and 18
27. d	2.0.0
28. c	6.6.2
29. a	8.0.0
30. b	6.4.0

*Please see the following page for math calculations.

MATH CALCULATIONS

17. Safe working load = diameter squared × 8 = 1 × 1 × 8 = 8
    Total weight of load = 70 tons
    70 tons ÷ 8 = 8.75
    In the Ratio Factors table, 8.98 is the next higher number under Ratio B. The number of line parts column gives 10 parts of line.

PERFORMANCE TASK WORKSHEET MATH CALCULATIONS

Answer to Task 6

1. S.W.L. = 1.25 × 1.25 × 8 = 12.5

   Total weight of load = 45 tons

   45 tons ÷ 12.5 = 3.6

   Refer to *Table 1*. From the column under Ratio A Bronze Bushed Sheaves, 4.39 is the next highest number over the calculated number of 3.6. The last column gives 5 parts of line.

Answer to Task 7

1. 78,750 lbs (gross capacity)
   −3,829 lbs (150-ton, 8-sheave hook block)
   −   180 lbs (rigging)
   74,741 lbs (net capacity)

2. 60,750 lbs (gross capacity)
   −3,829 lbs (150-ton, 8-sheave hook block)
   −8,412 lbs (extension retracted)
   −   200 lbs (rigging)
   48,309 lbs (net capacity)

3. 102,000 lbs (gross capacity)
   − 3,829 lbs (150-ton, 8-sheave hook block)
   −11,406 lbs (33 ft–58 ft extension extended)
   −   160 lbs (rigging)
   86,605 lbs (net capacity)

4. Boom tip height = 148.7
   15 × 2.34 = 35.1
   148.7 × 2.34 = 347.96 × 4 = 1391.84 + 35.10 = 1426.94 (round up to 1,427)

   30,500 lbs (gross weight)
   − 1,200 lbs (single-sheave boom point)
   −   865 lbs (15-ton hook block)
   − 2,065 lbs (100-ton hook block)
   −   345 lbs (rigging)
   − 1,427 lbs (rope)
   24,598 lbs (net capacity)

PERFORMANCE TASK WORKSHEET MATH CALCULATIONS

5. Boom tip height = 167.0
$30 \times 2.34 = 70.2$
$167 \times 2.34 = 390.78 \times 2 = 781.56$
$781.56 + 70.2 = 851.76$ (round up to 852)

16,300 lbs (gross capacity)
– 1,500 lbs (2-sheave upper boom point)
– 865 lbs (15-ton hook block)
– 2,065 lbs (100-ton block)
– 230 lbs (rigging)
– 852 lbs (rope)
10,788 lbs (net capacity)

6. Boom tip height = 171.1
$20 \times 2.34 = 46.8$
$171.1 \times 2.34 = 400.38 \times 4 = 1{,}601.5$
$1{,}602 + 46.8 = 1{,}648.8$ (rounded up to 1,649)

22,000 lbs (gross capacity)
– 2,065 lbs (100-ton block)
– 1,649 lbs (rope)
– 1,500 lbs (2-sheave boom point)
– 865 lbs 15-ton hook block
– 245 lbs (rigging)
15,676 lbs (net capacity)

7. 168,000 lbs (net capacity)
+ 2,065 lbs (100-ton block)
+ 875 lbs (rigging)
+ 2,306 lbs (100-ton block)
173,246 lbs (gross capacity)

34 ft radius is good for 173,246 lbs

8. 40,900 lbs (gross capacity)

9. 55.8° (minimum angle)

Telescopic Boom Attachment Assembly and Disassembly

21302-05

Trainee Name: ______________________________

Social Security Number: ____________________ **Date:** ____________

_____ 1. It is important to use the swing-away that is serial-number (or other identifier) specific to the crane for all of the following reasons *except* _____.

a. the load moment indicator (LMI) is only set for that particular extension
b. the crane will not function properly
c. damage may occur
d. there may be a clearance problem

_____ 2. During installation of the swing-away, what should you do after all boom sections have been fully retracted and the boom has been lowered to minimum elevation?

a. Guide the swing-away around until it latches to the main boom.
b. Pull the stop rail outward.
c. On the swing-away rear support bracket, pull the latch pin and swing the latch plate out of the way.
d. Rig either the main hoist or auxiliary hoist for single part of line.

_____ 3. Blocking should be used in a swing-away attachment _____.

a. at all times
b. if there are not enough people to help in the assembly
c. if there is not enough clearance and the swing-away needs to be kept above any obstacles
d. if there is difficulty in aligning the swing-away's attach points to the boom nose

_____ 4. During installation of an A-frame jib, what should be done after the hoist cable becket has been disconnected from the cable retainer pin in the A-frame nose?

a. Elevate the boom until the A-frame is parallel to the ground.
b. Install links in the correct fittings on each side of the boom nose applicable to the required degree of offset for the lift operation.
c. Remove the hoist cable from between the boom nose and the swing-away.
d. Retract and raise the boom 15 degrees so that the A-frame extends straight out.

_____ 5. When establishing the electrical connection from the main boom to the auxiliary single-sheave boom head, you must screw the _____ into the dummy socket on the auxiliary single-sheave boom head.

a. retaining pin
b. holding pin
c. short-circuit plug
d. circuit breaker

_____ 6. When assembling boom extension sections, always assemble the intermediate section with the cable support roller directly _____ of the boom head.

a. in front
b. to the left side
c. to the right side
d. to the rear

_____ 7. During assembly of the A-frame jib, what should be done before rigging the hoist cable?

a. Fully retract all boom sections and lower the boom to minimum elevation.
b. Reeve the hoist cable over the main boom sheave.
c. Reeve the hoist cable over the main boom sheave and swing-away sheave.
d. Position the cable guide in the vertical position so the cable passes through the guide.

_____ 8. During the removal of a swing-away extension, the operator should _____ immediately after fully retracting the telescopic section of the boom.

a. lower the main boom until the LMI switches off
b. override the LMI
c. wind in the hoist rope
d. turn the superstructure to the back or the side

_____ 9. The telescopic status and superstructure position when attaching boom extensions depend on the _____.

a. skill of the operator
b. mounted counterweight
c. experience of the assembly personnel
d. weight of the boom itself

_____ 10. The pin and retainer bushings should be removed from the base of the A-frame before reeling in the hoist cable because _____.

a. the hoist cable could be damaged
b. the pin and retainer bushings may get jammed in the boom head
c. damage to the A-frame could occur
d. it makes the disassembly process easier for personnel

_____ 11. Outriggers should _______ during boom assembly.

a. be extended halfway
b. be extended and set per manufacturer's recommendation
c. not be extended
d. be fully extended

_____ 12. Rope spacing is maximized in multi-part reeving so the traveling block _______.

a. will not two-block
b. hangs straight and level
c. can withstand the weight of the load
d. has sufficient movement capabilities

_____ 13. During installation of the A-frame jib, what should be done after the hoist cable has been reeled out and the hoist cable becket has been disconnected from the mast lug?

a. Reel in the hoist cable and attach the becket to the lug on the mast.
b. Reeve the hoist cable over the boom nose center idler sheave, swing-away sheave, and A-frame sheave.
c. Set the required jib offset by adjusting the rear support cable turnbuckles.
d. Raise the cable roller to an upright position so the hoist cable is under the sheave.

_____ 14. The boom must be raised _____ to prevent the swing-away from swinging around after removing the attach pins from the front support bracket.

a. 10 degrees
b. 15 degrees
c. 20 degrees
d. to horizontal

_____ 15. The _____ must be moved into working position according to the correct inclination after the swing-away has been locked to the main boom.

a. cable roller sheave
b. hoist cable
c. load block
d. attachment pins

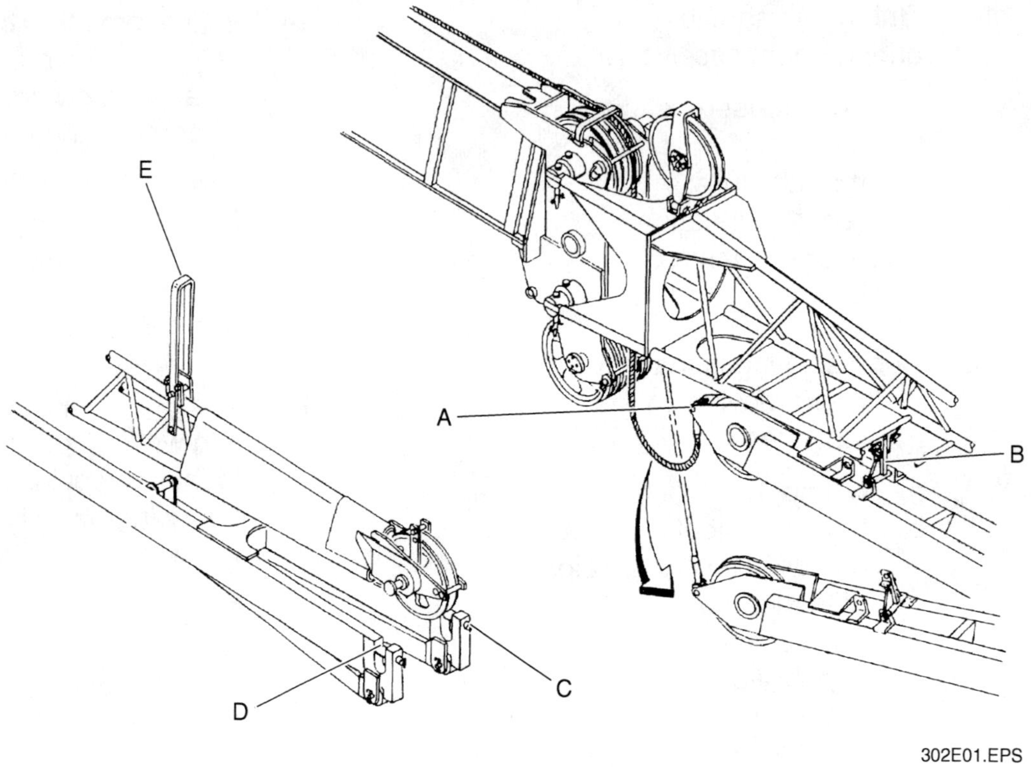

Figure 1

Using *Figure 1*, match the letters on the figure to the correct descriptions.

_____ 16. Retainer bushing

_____ 17. Cable-retaining pin

_____ 18. Retainer pin

_____ 19. Attaching link

Using *Figure 2*, match the letters on the figure to the correct descriptions.

_____ 20. Cotter pin

_____ 21. Hitch pin

_____ 22. Manual load pin

_____ 23. Section 2

_____ 24. Section 3

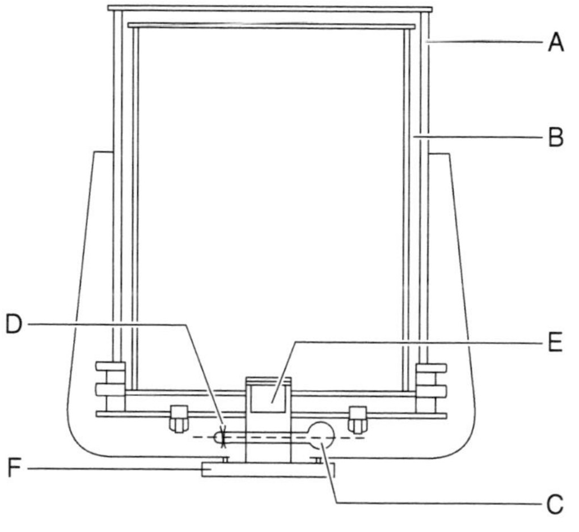

Figure 2

_____ 25. When removing intermediate boom sections as discussed in this module, the operator must lower the main boom until _____.

a. the boom extension is in the 45-degree offset position
b. the manual extension pin hole is aligned with the access section
c. the LMI triggers a shutdown
d. section 3 is pinned to the lower telescope cylinder

_____ 26. A fly is another name for a(n) _____.

a. A-frame jib
b. swing-away lattice extension
c. intermediate boom section
d. single-sheave boom head

_____ 27. A link or sleeve used as a means of coupling and tightening two parts is a _____.

a. turnbuckle
b. pin
c. mast
d. bushing

_____ 28. A projecting piece by which something is held or supported is a _____.

a. mast
b. pin
c. lug
d. bushing

_____ 29. A wedge socket used as an end preparation for wire rope is a(n) _____.

a. bushing
b. idler
c. becket
d. lug

_____ 30. A(n) _____ consists of two side rails and cross members that are constructed of rectangular tubing welded together.

a. power-pinned fly
b. A-frame jib
c. swing-away lattice extension
d. mast

notes

Craft: Mobile Crane Operations

Module Number: 21302-05

Module Title: Telescopic Boom Attachment Assembly and Disassembly

contren Learning Series

TRAINEE NAME: ____________________

TRAINEE SOCIAL SECURITY NUMBER: ____________________

CLASS: ____________________

TRAINING PROGRAM SPONSOR: ____________________

____

INSTRUCTOR: ____________________

Rating Levels: (1) Passed: performed task (2) Failed: did not perform task
Also, list the date the testing for each task was completed.

Recognition: When testing for the NCCER Standardized Craft Training Program, be sure to record Performance testing results on Craft Training Report Form 200, and submit the results to the Training Program Sponsor.

Objective	TASK	RATING
1	1. Analyze the site for assembly/disassembly room.	
2	2. Analyze and determine counterweight requirements needed for boom assembly.	
1	3. Determine the ground stability and possible ground support needed for assembly and disassembly.	

continued

Craft: Mobile Crane Operations

Module Number: 21302-05

Module Title: Telescopic Boom Attachment Assembly and Disassembly

contren Learning Series

Objective	TASK	RATING
3	4. Assemble and disassemble crane attachments and components.	
3	5. Install and remove an auxiliary single-sheave boom head.	
3	6. Properly install wire rope and all lifting attachments.	

ANSWER KEY

Answer	Section Reference
1. d	4.0.0
2. d	4.1.0
3. d	4.1.0
4. c	5.1.0
5. c	8.0.0
6. a	6.1.0
7. a	5.1.0
8. d	4.2.0
9. b	3.0.0
10. c	5.2.0
11. b	4.1.0
12. b	9.0.0
13. c	5.1.0
14. d	4.1.0
15. a	4.1.0
16. d	5.2.0; Figure 8
17. a	5.2.0; Figure 8
18. c	5.2.0; Figure 8
19. b	5.2.0; Figure 8
20. d	7.1.0; Figure 12
21. c	7.1.0; Figure 12
22. e	7.1.0; Figure 12
23. a	7.1.0; Figure 12
24. b	7.1.0; Figure 12
25. c	6.2.0
26. b	4.0.0
27. a	5.1.0; Glossary
28. c	5.1.0; Glossary
29. c	5.1.0; Glossary
30. b	5.0.0

notes

Advanced Operational Techniques

21303-05

Trainee Name: ______________________________

Social Security Number: ______________________ **Date:** ____________

_____ 1. In which situation would personnel be lifted by a crane or derrick?

a. Other means would be more hazardous or not possible.
b. There are no other exceptions.
c. The operator making the lift is licensed.
d. The personnel are not to be lifted above 30 feet.

_____ 2. Multiple-crane lifts should always be considered _____ lifts.

a. static
b. routine
c. critical
d. blind

_____ 3. In multiple-crane lifts, no crane should be loaded more than _____ percent of its net capacity.

a. 60
b. 65
c. 70
d. 75

_____ 4. When the load being picked up or set down is *not* visible to the operator, the lift is a(n) _____ lift.

a. signalled
b. blind
c. prohibited
d. auxiliary

_____ 5. The safe limit for conducting lifting operations in windy conditions is _____.

a. that which is recommended by the manufacturer
b. 10 mph
c. 15 mph
d. 25 mph

_____ 6. On lifts from the main boom, required capacity is determined by _____.

a. the boom angle indicator
b. load radius
c. the angle of the main boom
d. load weight

_____ 7. In a critical lift situation, unless the crane sits on a _____ foundation, outrigger blocking must be used.

a. wood
b. solid dirt
c. sufficient concrete
d. barge

_____ 8. Prior to lifting near a transmitter, _____ must be removed from the work area.

a. electrical grounds
b. ground jumper cables
c. combustible materials
d. alligator clips

_____ 9. During wrecking ball operations, _____ loading is imposed on the boom structure and other parts of the crane by the movement and impact of the ball.

a. horizontal
b. static
c. dynamic
d. sway

_____ 10. At the intended point of impact during vertical-drop balling, there should be sufficient _____ to allow the ball to be brought to rest.

a. boom length
b. boom angle
c. load radius
d. length of rope left on the crane

_____ 11. When demolition is accomplished by swinging the ball in line with the boom, _____ is used to pull the ball in towards the crane.

a. a tagline handler
b. rope from second drum on the crane
c. additional rope on the drum of the crane
d. personnel in a standby aerial work platform bucket

_____ 12. When using a vacuum lifting device, the operator should lift the load a few inches and stop before proceeding in order to _____.

a. make sure the rigging is not getting tangled
b. make sure the supervisor is aware that a lifting operation is taking place
c. make sure the device has been correctly applied and that a stable vacuum level exists
d. check the stress on the rigging

_____ 13. Vacuum lifts must be constructed so that the rated load is *no* greater than _____ percent of the ultimate pad capacity for horizontal surface lifts.

a. 30
b. 40
c. 50
d. 60

_____ 14. When performing a load test on a vacuum lifting device, the load rating should *not* be more than _____ percent of the maximum load sustained during the test.

a. 50
b. 60
c. 70
d. 80

_____ 15. The _____, the more lines of magnetic force are able to flow.

a. longer the load
b. thicker the load
c. larger the magnet
d. wider the magnet

_____ 16. On a magnet's surface, paper, dirt, rust, paint, and scale have the same effect as _____.

a. acid
b. water
c. gas
d. air

_____ 17. You should *never* disconnect the magnet from its power source while it is energized because _____ will occur.

a. static loading
b. electrical arcing
c. slewing
d. duty cycle overload

_____ 18. In cold weather operations, the crane's engine should be kept at average speed until the hydraulic oil has reached _____.

a. 30°F
b. 35°F
c. 40°F
d. 50°F

_____ 19. When warming up a truck crane in cold weather, the engine should *not* be raced because _____.

a. it will overheat
b. cavitation in the hydraulic pumps may occur
c. the pistons may stick
d. the gauges will give inaccurate readings because the engine is being shocked

_____ 20. How should outrigger operation be conducted in cold weather operations?

a. They should not be extended.
b. Extend them quickly so there is no opportunity for them to get stuck.
c. Extend them only halfway so as not to shock the cylinders.
d. Slowly extend them six inches and retract, extend halfway and retract, and then extend full stroke.

_____ 21. For an electromagnet, the _____ the lines of magnetic force and the _____ the ferrous load, the greater the effectiveness of the magnet.

a. greater; thinner
b. lesser; thinner
c. greater; thicker
d. lesser; thicker

_____ 22. Personnel lifts are always classified as _____ lifts.

a. critical
b. blind
c. prohibited
d. auxiliary

_____ 23. A(n) _____ brake for a hoist drum eliminates overspooling problems.

a. electric
b. disc
c. friction band
d. hydraulic

_____ 24. Potential damage to the swing-circle bearing is eliminated by _____.

a. front and rear pendants
b. hard mounting of the upperworks
c. a special drum brake
d. rapid cycling

_____ 25. Critical lifts plans always require a(n) _____.

a. verbal approval by qualified personnel
b. written approval by qualified personnel
c. operator approval only
d. supervisor approval only

_____ 26. An electromagnet can lift _____.

a. stainless steel
b. copper
c. aluminum
d. low-carbon steel

_____ 27. Slew demolition _____.

a. is a recommended method of demolition
b. is not a recommended method of demolition
c. should be used instead of vertical drop demolition
d. is an alternate compaction demolition

_____ 28. The demolition method that produces the least strain on a crane is _____ demolition.

a. vertical-drop
b. in-line
c. slew
d. radial

_____ 29. Peel is a condition associated with _____.

a. in-line demolition
b. slew demolition
c. electromagnet lifting
d. dynamic compaction

_____ 30. Besides being used for initiation of in-line demolition, a pulling rope can help _____.

a. keep a wrecking ball from dropping too far
b. keep a wrecking ball from exceeding the operating radius of the crane
c. retrieve a wrecking ball
d. in the cracking and demolition process

Craft: Mobile Crane Operations

Module Number: 21303-05

Module Title: Advanced Operational Techniques

contren Learning Series

TRAINEE NAME: ______________________________

TRAINEE SOCIAL SECURITY NUMBER: ______________________________

CLASS: ______________________________

TRAINING PROGRAM SPONSOR: ______________________________

____

INSTRUCTOR: ______________________________

Rating Levels: (1) Passed: performed task (2) Failed: did not perform task
Also, list the date the testing for each task was completed.

Recognition: When testing for the NCCER Standardized Craft Training Program, be sure to record Performance testing results on Craft Training Report Form 200, and submit the results to the Training Program Sponsor.

Objective	TASK	RATING
2	1. Perform a blind lift.	
8	2. Identify the components of a crane that should be observed closely before cold weather operation.	

notes

ANSWER KEY

Answer	Section Reference
1. a	2.0.0
2. c	3.0.0
3. d	3.0.0
4. b	4.0.0
5. a	5.0.0
6. b	5.0.0
7. c	5.0.0
8. c	6.0.0
9. c	7.0.0
10. d	7.1.0
11. b	7.2.0
12. c	8.0.0
13. c	8.0.0
14. d	8.0.0
15. b	9.1.0
16. d	9.3.0
17. b	9.4.0
18. d	10.0.0
19. b	10.0.0
20. d	10.0.0
21. c	9.1.0
22. a	2.0.0
23. b	7.1.0
24. b	7.1.0
25. b	3.0.0
26. d	9.2.0
27. b	7.3.0
28. b	7.2.0
29. c	9.3.0
30. b	7.2.0

notes

Lift Planning

21304-05

Trainee Name: ____________________

Social Security Number: ____________________ **Date:** ____________

_____ 1. When is a new lift plan required?

a. Once a lift plan is made out, a new plan is never required.
b. If there is a change in crane operator.
c. If there is any deviation from the original lift plan.
d. If a different signal person is used.

_____ 2. Electrical hazards in the lift area and swing direction and degree are factors to consider regarding the _____ portion of a lift plan.

a. weight
b. slings/sizing
c. cable
d. crane placement

_____ 3. When using two cranes to lift an object from each crane's main boom, the location of each crane's _____ above the load has to be chosen carefully so the load is shared between both cranes.

a. swing-away extension
b. hook block
c. erected jib
d. auxiliary boom sheave

_____ 4. When planning a lift involving two cranes, calculate the lowest net capacities that both cranes will have during the lift. The next step is to _____.

a. locate the center of gravity of both cranes
b. make sure that the sum of the lowest rated net capacities of the cranes exceeds the load weight
c. position the load blocks relative to the load's center of gravity
d. identify what kind of ground support is required

_____ 5. A lift plan should be discussed _____.

a. when the operator is about to start the lift
b. at a pre-lift meeting where all personnel involved in the lift are present
c. at a meeting involving the site superintendent, operator, and the owner of the crane
d. if there are problems during the lift

_____ 6. In a multiple-crane lift using an equalizer, the weight of the equalizer is considered part of the _____.

a. deductions from gross capacity
b. total gross capacities of Cranes One and Two
c. total net capacities of Cranes One and Two
d. load weight

_____ 7. In a multiple-crane lift using an equalizer and lifting from the main booms, measure the _____ of each crane to its pick-up point on the equalizer.

a. center of gravity
b. load radius
c. boom point elevation
d. boom length

_____ 8. Equipment condition is categorized under _____ in a lift plan.

a. pre-lift checklist
b. weight
c. rope
d. sizing and type of slings

_____ 9. The person in charge of planning a critical lift is the _____ or other qualified person.

a. rigging superintendent
b. crane operator
c. client representative
d. signal person

_____ 10. During a critical lift, only _____ may give signals to the crane operator.

a. the project manager
b. the safety representative
c. the rigging superintendent
d. designated, qualified signalers

_____ 11. Sling lengths need to be checked to ensure the horizontal angle is greater than _____ degrees.

a. 30
b. 40
c. 50
d. 60

_____ 12. The object to be lifted, its dimensions and general configuration, and its gross weight are _____ engineering requirements.

a. location
b. operating radius
c. load data
d. lift attachment locations

_____ 13. On multiple-crane lifts, the load on each crane should *not* exceed _____ percent of its capacity.

a. 55
b. 65
c. 75
d. 85

_____ 14. The net capacity of Crane One is 130,000 lbs. The net capacity of Crane Two is 90,000 lbs. The weight of the load is 145,000 lbs. The weight of the equalizer beam is 3,000 lbs. The distance between the hook blocks is 25 feet. The load's pick up point on the equalizer beam is _____.

a. 7 feet from Crane One and 15 feet from Crane Two
b. 8 feet from Crane One and 15 feet from Crane Two
c. 9 feet from Crane One and 15 feet from Crane Two
d. 10 feet from Crane One and 15 feet from Crane Two

_____ 15. If the gross capacity of a crane is 75,000 lbs and the total weight of deductions is 4,500 lbs, the net capacity of the crane in this configuration is _____.

a. 52,875 lbs
b. 70,500 lbs
c. 79,500 lbs
d. impossible to calculate without more information

_____ 16. If the net capacity of a crane in a particular configuration is 85,000 lbs and the total weight of deductions is 6,400 lbs, the gross capacity is _____.

a. 78,600 lbs
b. 91,400 lbs
c. 95,600 lbs
d. impossible to calculate without more information

_____ 17. When lifting from a jib of a lattice boom crane, the line to the jib top is considered part of the effective weight of _____.

a. the load to be lifted, not including rigging
b. the crane components below the lifting point
c. the crane components past the lifting point
d. any additional crane components and/or boom components

_____ 18. The weight of rigging on the load block is considered part of the weight of _____.

a. the hook block
b. unaccounted material
c. crane components past the lifting point
d. crane components below the lifting point

_____ 19. Which of the following is an example of a critical lift?

a. A lift requiring calculations of net capacity.
b. A lift using a heavy-lift beam.
c. A lift that requires a spreader.
d. A lift requiring exceptional care in handling.

_____ 20. When using an equalizer beam, ensure that the sum of the lowest rated net capacities of the cranes _____.

a. exceeds the load weight plus the equalizer and sling weight
b. does not exceed the load weight plus the equalizer and sling weight
c. is less than 75 percent of the load weight
d. exceeds the load weight minus the equalizer and sling weight

_____ 21. If the gross capacity of a crane is 120,000 lbs and the total weight of deductions is 20,000 lbs, the net capacity of the crane in this configuration is _____.

a. 100,000 lbs
b. 120,000 lbs
c. 140,000 lbs
d. impossible to calculate without more information

_____ 22. In a multiple-crane lift, if the total net capacity of Crane One is 100,000 lbs and the total net capacity of Crane Two is 90,000 lbs, the total load weight must be *no* more than _____ lbs to meet the 75 percent of capacity guidelines.

a. 135,000
b. 142,500
c. 163,500
d. 190,000

_____ 23. In a multiple-crane lift, if the total net capacity of Crane One is 100,000 lbs and the total load weight is 120,000 lbs, the total net capacity of Crane Two must be at least _____ lbs to meet the 75 percent of capacity guidelines.

a. 20,000
b. 50,000
c. 60,000
d. 80,000

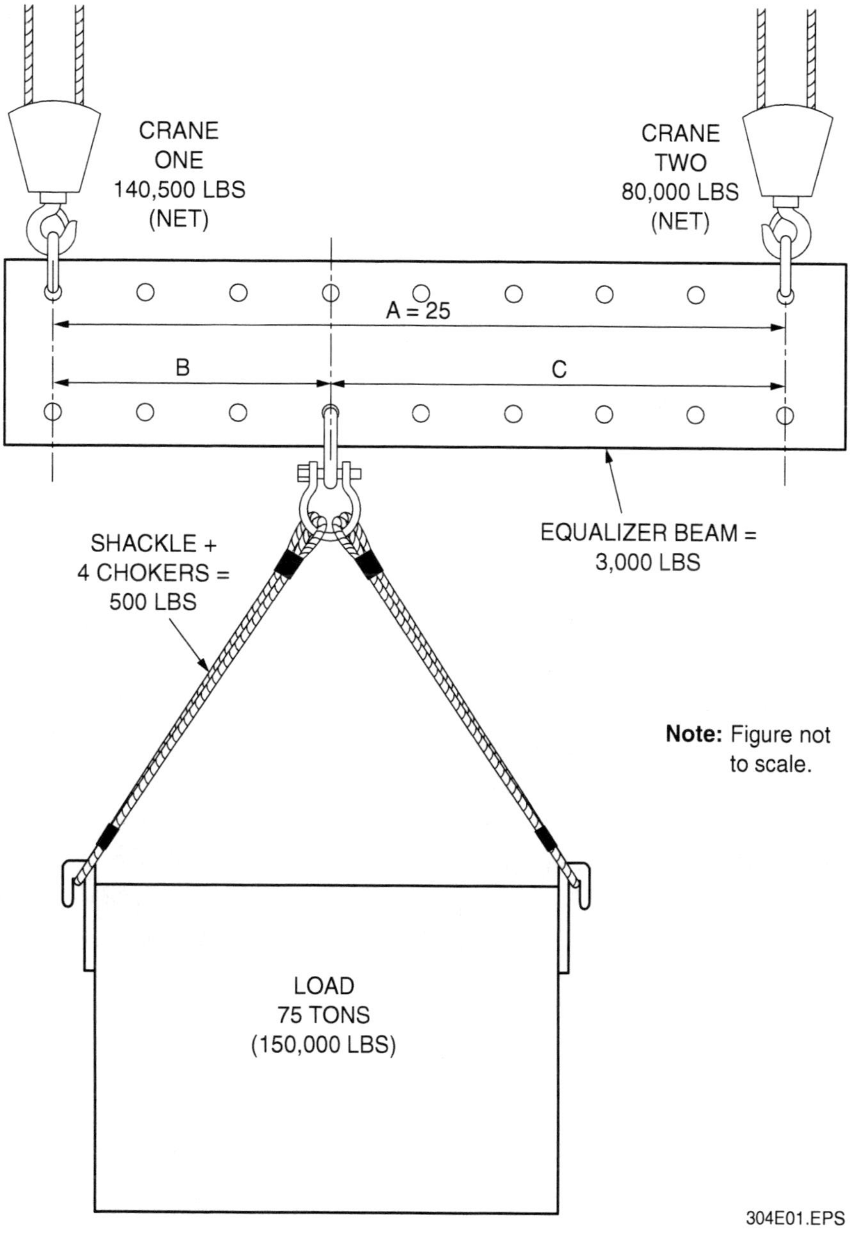

Figure 1

_____ 24. What is the total load weight in *Figure 1*?

a. 153,000 lbs
b. 153,500 lbs
c. 75 tons
d. impossible to calculate without more information

_____ 25. What percentage of total net capacity is the load in *Figure 1*?

a. 50
b. 60
c. 70
d. impossible to calculate without more information

_____ 26. In *Figure 1,* if A = 33 feet, then what is the value for B?

a. 10
b. 12
c. 15
d. 20

_____ 27. In *Figure 1,* if C = 20 feet, then what is the actual load on Crane One?

a. 67,333 lbs
b. 76,500 lbs
c. 93,030 lbs
d. 153,000 lbs

_____ 28. The formula C/A × (load weight) can be used to determine the _____.

a. load on Crane One
b. attachment point for the equalizer bar
c. percentage of the load carried by Crane One
d. net capacity of the two cranes

_____ 29. The formula Net Capacity of Crane Two/Total Net Capacity of Cranes One and Two × A can be used to determine the _____.

a. load on Crane Two
b. attachment point of the load on the equalizer beam
c. percentage of the load carried by Crane Two
d. net capacity of Crane Two

_____ 30. Given weights of 1,500 lbs for a load block, 850 lbs for a hook block, 400 lbs for rigging, and 1,825 lbs for rope, what is the net capacity of a crane when lifting from the main boom if the gross capacity is 89,000 lbs?

a. 84,405 lbs
b. 84,425 lbs
c. 86,125 lbs
d. 93,575 lbs

Craft: Mobile Crane Operations

Module Number: 21304-05

Module Title: Lift Planning

contren® Learning Series

TRAINEE NAME: ______________________________

TRAINEE SOCIAL SECURITY NUMBER: ______________________________

CLASS: ______________________________

TRAINING PROGRAM SPONSOR: ______________________________

____

INSTRUCTOR: ______________________________

Rating Levels: (1) Passed: performed task (2) Failed: did not perform task
Also, list the date the testing for each task was completed.

Recognition: When testing for the NCCER Standardized Craft Training Program, be sure to record Performance testing results on Craft Training Report Form 200, and submit the results to the Training Program Sponsor.

Objective	TASK	RATING
1	1. Fill out a lift plan based on a given scenario.	
3	2. Perform single-crane lifting calculations.	
3	3. Perform multiple-crane lifting calculations.	

notes

Trainee Name: ____________________

Social Security Number: ____________________ Date: ____________

Task 2 – Perform single-crane lifting calculations.

For Question 1, refer to the Grove RT750 load/capacity charts shown in *Figures 1* through *4*.

__________ 1. What is the net capacity when lifting from the main boom with the RT750 configured as follows?

- On outriggers, 360 degrees, fully extended
- Boom length of 80 ft
- Load radius of 15 ft
- 45-ton, 3-sheave load block without cheekplates reeved with required parts of line
- Auxiliary boom head installed
- 10-ton headache ball on auxiliary line
- Rigging 100 lbs

Rated Lifting Capacities in Pounds (35 Ft – 110 Ft Boom) On Outriggers Fully Extended – 360°

Radius in Feet	#0001								
	Main Boom Length in Feet								
	35	40	50	*60	70	80	90	100	110
10	100,000 (63.5)	80,400 (66.5)	74,400 (71.5)	44,600 (75.5)					
12	88,050 (60)	79,050 (63.5)	70,900 (69)	44,600 (74)	@35,600 (75.5)				
15	74,500 (54)	67,450 (59)	63,350 (65.5)	44,600 (71)	35,600 (74)	@33,000 (75.5)			
20	54,700 (43)	53,850 (50.5)	50,900 (59)	44,600 (66)	35,600 (70)	33,000 (72.5)	25,500 (75)	@23,300 (75.5)	
25	41,450 (29)	41,150 (40.5)	40,700 (52.5)	40,350 (60.5)	35,550 (65.5)	33,000 (69)	25,500 (71.5)	23,300 (74)	@18,500 (75.5)
30		32,450 (28)	32,050 (45)	31,750 (55)	30,550 (61)	28,950 (65)	25,500 (68)	23,300 (71)	18,500 (73)
35			25,950 (36.5)	25,650 (48.5)	26,500 (56.5)	24,900 (61)	23,000 (64.5)	21,200 (68)	18,500 (70.5)
40	See Note 16		21,400 (25)	21,150 (41.5)	22,000 (51.5)	21,750 (57)	20,000 (61)	18,450 (65)	18,000 (67.5)
45				17,600 (33.5)	18,500 (46)	19,100 (53)	17,600 (57.5)	16,300 (61.5)	15,750 (65)
50				14,600 (23)	15,250 (39.5)	15,700 (48)	15,650 (53.5)	14,400 (58)	13,950 (62)
55					12,650 (32.5)	13,100 (43)	13,550 (49.5)	12,850 (54.5)	12,450 (59)
60					10,500 (23)	11,000 (37.5)	11,450 (45)	11,550 (51)	11,150 (55.5)
65						9,350 (31)	9,780 (40.5)	10,200 (47)	10,050 (52.5)
70						7,870 (22)	8,370 (35)	8,780 (43)	9,090 (49)
75							7,180 (28.5)	7,590 (38.5)	7,980 (45)
80							6,120 (20)	6,560 (33)	6,950 (41)
85								5,680 (27)	6,060 (37)
90								4,910 (19)	5,280 (32)
95									4,600 (26)
100									3,990 (18.5)
Minimum boom angle (deg.) for indicated length									0
Maximum boom length (ft.) at 0 deg. boom angle (no load)									110

Note: ( ) Boom angles are in degrees.
#LMI operating code. Refer to LMI manual for instructions.
@This capacity is based upon maximum boom angle.
*60 ft. boom length is with inner-mid extended and outer-mid & fly retracted.

304E02.EPS

Figure 1

Rated Lifting Capacities in Pounds
35' – 110' Boom
On Outriggers 0% Extended (10'–5" Spread) – 360°

Radius in Feet	#8001 Main Boom Length in Feet								
	35	40	50	*60	70	80	90	100	110
10	79,850 (63.5)	75,500 (66.5)	68,400 (71.5)	44,600 (75.5)					
12	60,500 (60)	57,550 (63.5)	52,750 (69)	44,600 (74)	@35,600 (75.5)				
15	41,750 (54)	41,550 (59)	38,400 (65.5)	35,800 (71)	35,350 (74)	@33,000 (75.5)			
20	25,250 (43)	25,000 (50.5)	24,450 (59)	23,550 (66)	23,750 (70)	23,700 (72.5)	23,500 (75)	@23,300 (75.5)	
25	17,100 (29)	16,800 (40.5)	16,200 (52.5)	15,600 (60.5)	16,400 (65.5)	17,150 (69)	17,250 (71.5)	17,200 (74)	@17,100 (75.5)
30		11,900 (28)	11,250 (45)	10,600 (55)	11,350 (61)	12,050 (65)	12,750 (68)	13,200 (71)	13,200 (73)
35			7,990 (36.5)	7,330 (48.5)	8,020 (56.5)	8,680 (61)	9,340 (64.5)	9,980 (68)	10,400 (70.5)
40	See Note 16		5,660 (25)	4,970 (41.5)	5,630 (51.5)	6,260 (57)	6,880 (61)	7,500 (65)	8,100 (67.5)
45				3,200 (33.5)	3,840 (46)	4,450 (53)	5,050 (57.5)	5,640 (61.5)	6,220 (65)
50				1,830 (23)	2,450 (39.5)	3,040 (48)	3,620 (53.5)	4,200 (58)	4,760 (62)
55					1,340 (32.5)	1,910 (43)	2,480 (49.5)	3,040 (54.5)	3,590 (59)
60							1,550 (45)	2,100 (51)	2,630 (55.5)
65								1,310 (47)	1,840 (52.5)
70									1,160 (49)
0.1A (lbs.)	765	805	870	915	840	790	750	715	690
Minimum boom angle (deg.) for indicated length				5	26	38	42	45	48
Maximum boom length (ft.) at 0 deg. boom angle (no load)				50					

Note: ( ) Boom angles are in degrees.
#LMI operating code. Refer to LMI manual for instructions.
@This capacity is based upon maximum boom angle.
*60 ft. boom length is with inner-mid extended and outer-mid & fly retracted.

Lifting Capacities at Zero Degree Boom Angle on Outriggers 0% Extended - 360°									
Boom Angle	Main Boom Length in Feet								
	35	40	50						
0°	12,900 (29.2)	9,100 (34.3)	4,140 (44.3)						

Note: ( ) Reference radii in feet.

304E03.EPS

Figure 2

ON RUBBER CAPACITIES WITH 29.5 × 25 TIRES
STATIONARY CAPACITIES - 360°

Radius in Feet	#9005 Main Boom Length in Feet				
	35	40	50	*60	70
10	47,400 (63.5)	38,900 (66.5)	30,550 (71.5)		
12	37,650 (60)	37,650 (63.5)	30,550 (69)	22,250 (74)	
15	26,300 (54)	26,300 (59)	26,300 (65.5)	17,300 (71)	17,300 (74)
20	15,800 (43)	15,800 (50.5)	15,800 (59)	15,800 (66)	15,800 (70)
25	10,350 (29)	10,350 (40.5)	10,350 (52.5)	10,350 (60.5)	10,350 (65.5)
30		7,020 (28)	7,020 (45)	7,020 (55)	7,020 (61)
35			4,750 (36.5)	4,750 (48.5)	4,750 (56.5)
40			3,120 (25)	3,120 (41.5)	3,120 (51.5)
45				1,880 (33.5)	1,880 (46)

#LMI operating code. Refer to LMI manual for instructions.
*60 ft. boom length is with inner-mid extended and outer-mid & fly retracted.

STATIONARY CAPACITIES - DEFINED ARC OVER FRONT (See Note 3)

Radius in Feet	#9005 Main Boom Length in Feet				
	35	40	50	*60	70
10	47,400 (63.5)	38,900 (66.5)	30,550 (71.5)		
12	38,900 (60)	38,900 (63.5)	30,550 (69)	22,250 (74)	
15	30,550 (54)	30,550 (59)	26,300 (65.5)	17,300 (71)	17,300 (74)
20	26,050 (43)	26,050 (50.5)	22,250 (59)	17,300 (66)	17,300 (70)
25	21,150 (29)	21,150 (40.5)	21,150 (52.5)	17,300 (60.5)	17,300 (65.5)
30		15,300 (28)	15,300 (45)	15,300 (55)	15,300 (61)
35			11,350 (36.5)	11,350 (48.5)	11,350 (56.5)
40			8,590 (25)	8,590 (41.5)	8,590 (51.5)
45				6,490 (33.5)	6,490 (46)
50				4,860 (23)	4,860 (39.5)
55					3,630 (32.5)
60					2,610 (23)

#LMI operating code. Refer to LMI manual for instructions.
*60 ft. boom length is with inner-mid extended and outer-mid & fly retracted.

304E04.EPS

Figure 3

LINE PULLS AND REEVING INFORMATION

HOISTS	CABLE SPECS.	PERMISSIBLE LINE PULLS	NOMINAL CABLE LENGTH
Main & Aux. Model 30	¾" (19 mm) 18×19 Class or 35×7 Rotation Resistant Min. Breaking Str. 64,600 lbs.	12,920 lbs.	500 ft.

WEIGHT REDUCTIONS FOR LOAD HANDLING DEVICES

32 ft. Fixed Extension (Stowed on Boom Base Section)	521 lbs.
32 ft. – 56 ft. Tele. Extension (Stowed on Boom Base Section)	718 lbs.
32 ft. Fixed Extension (Erected)	5,851 lbs.
32 ft. Tele. Extension (Erected)	8,636 lbs.
44 ft. Tele. Extension (Erected)	10,307 lbs.
56 ft. Tele. Extension (Erected)	12,066 lbs.

* Reduction of main boom capacities

AUXILIARY BOOM HEAD	220 lbs.
HOOKBLOCKS and HEADACHE BALLS:	
50 Ton, 4 Sheave	1,469 lbs. +
45 Ton, 3 Sheave w/ Cheekplates	977 lbs. +
45 Ton, 3 Sheave w/o Cheekplates	830 lbs. +
15 Ton, 1 Sheave	420 lbs. +
5 Ton Headache Ball	172 lbs. +
7½ Ton Headache Ball	338 lbs. +
10 Ton Headache Ball	560 lbs. +

+ Refer to rating plate for actual weight.

When lifting over swingaway and/or jib combinations, deduct total weight of all load handling devices reeved over main boom nose directly from swingaway or jib capacity.

NOTE: All load handling devices and boom attachments are considered part of the load and suitable allowances MUST BE MADE for their combined weights. Weights are for Grove furnished equipment.

304E05.EPS

Figure 4

notes

Trainee Name: ______________________________

Social Security Number: ______________________ Date: ______________

Task 3 – Perform multiple-crane lifting calculations.

_________ 1. Determine if the following lift can be accomplished without the share of the load exceeding 75 percent of either crane's capacity. (The position of the load blocks have already been given.)

- Net capacity Crane One = 128,600 lbs
- Net capacity Crane Two = 94,500 lbs
- Total combined capacity = 223,100 lbs
- Load weight = 162,000 lbs
- Dimension A = 34 ft for Crane One to the load center of gravity
- Dimension B = 46.2 ft for Crane Two to the load center of gravity

notes

ANSWER KEY

Answer	Section Reference
1. c	1.0.0
2. d	2.0.0
3. b	3.2.0
4. b	3.2.0
5. b	6.0.0
6. d	3.2.0
7. b	3.2.0
8. b	2.0.0
9. a	4.0.0
10. d	4.0.0
11. d	5.0.0
12. c	5.0.0
13. c	3.2.0
14. d*	3.2.0
15. b*	3.1.0
16. b*	3.1.0
17. c	3.0.0
18. d	3.0.0
19. d	4.0.0
20. a	3.2.0
21. a*	3.1.0
22. b*	3.2.0
23. c*	3.2.0
24. b*	3.1.0
25. c*	3.1.0
26. b*	3.2.0
27. c*	3.2.0
28. a	3.2.0
29. b	3.2.0
30. b*	3.1.0

*Please see the following pages for math calculations.

MATH CALCULATIONS

14. B = (net capacity of Crane Two/net capacity of Cranes One and Two) × A
    A = 25 feet
    B = (90,000/220,000) × 25
    B = 0.41 × 25
    B = 10 feet
    C = A – B
    C = 15 feet

15. Net capacity = gross capacity – deductions
    Net capacity = 75,000 – 4,500
    Net capacity = 70,500 lbs

16. Gross capacity = net capacity + deductions
    Gross capacity = 85,000 + 6,400
    Gross capacity = 91,400

21. Net capacity = gross capacity – deductions
    Net capacity = 120,000 – 20,000
    Net capacity = 100,000 lbs

22. Load weight = 75 percent × (net capacity of Crane One + net capacity of Crane Two)
    Load weight = 0.75(100,000 + 90,000)
    Load weight = 142,500 lbs

23. Load Weight = 75 percent × (net capacity of Crane One + net capacity of Crane Two)
    120,000 = 0.75 (100,000 + net capacity of Crane Two)
    120,000 = 75,000 + 0.75 (net capacity of Crane Two)
    45,000 = 0.75 (net capacity of Crane Two)
    45,000/0.75 = net capacity of Crane Two
    60,000 lbs = net capacity of Crane Two

24. Total load weight = load + equalizer + sling weight
    Total load weight = 150,000 + 3,000 + 500
    Total load weight = 153,500 lbs

MATH CALCULATIONS

25. Percentage of total net capacity = total load/(net capacity Crane One + net capacity Crane Two)
    Percentage of total net capacity = 153,500/(140,000 + 80,000)
    Percentage of total net capacity = 153,500/220,000
    Percentage of total net capacity = 0.697 or 70 percent

26. B = (net capacity Crane Two/ total net capacity of Cranes One and Two) $\times$ A
    B = (80,000/220,000) $\times$ 33
    B = 12

27. Load on Crane One = C/A $\times$ (total load weight)
    Load on Crane One = 20/33 $\times$ 153,500
    Load on Crane One = 93,030 lbs

30. 89,000 – (1,500 + 850 + 400 + 1,825) = 84,425

notes

PERFORMANCE TASK WORKSHEET MATH CALCULATIONS

Answer to Task 2

1. 33,000 lbs (gross capacity)
   – 830 lbs (45-ton, 3-sheave load block w/o cheekplates)
   – 220 lbs (auxiliary head)
   – 560 lbs (headache ball)
   – 100 lbs (rigging)
   31,290 lbs (net capacity)

Answer to Task 3

1. Load Crane One = [B ÷ (A + B)] × load weight
   Load Crane One = [46.2 ÷ (34 + 46.2)] × 162,000
   Load Crane One = 93,322 lbs
   Crane One capacity = 128,600 × 0.75 = 96,450 lbs
   Crane One has a safe lift.

   Load Crane Two = [34 ÷ (34 + 46.2)] × 162,000
   Load Crane Two = 68,678 lbs
   Crane Two Capacity = 94,500 lbs × 0.75 = 70,875 lbs
   Crane Two has a safe lift.

notes

Hoisting Personnel

21305-05

Trainee Name: ______________________________

Social Security Number: ____________________ **Date:** ____________

_____ 1. The personnel platform's rigging may only be used for the platform, employees, tools, and _____.

a. extremely heavy loads
b. tilt-up walls
c. the crane operator
d. materials

_____ 2. For hoisting operations, the crane must be level within _____ percent, with outriggers fully deployed per manufacturer's specifications.

a. 1
b. 2
c. 3
d. 4

_____ 3. During proof testing, the platform and rigging are required to be loaded to _____ percent of the platform's rated capacity.

a. 100
b. 110
c. 115
d. 125

_____ 4. According to *OSHA 1926.104*, lifelines must be secured above the point of operation to an anchorage or structural member capable of supporting a minimum dead weight of _____ pounds.

a. 5,400
b. 6,400
c. 7,400
d. 8,400

_____ 5. Under ASME guidelines, platforms must have a minimum platform rating of _____ pounds.

a. 300
b. 400
c. 500
d. 600

_____ 6. According to *OSHA 1926.550(g)*, all eyes in wire rope slings must be fabricated with _____.

a. beam clamps
b. anchors
c. bolts
d. thimbles

_____ 7. Under both OSHA and ASME guidelines, the platform must be held _____ minutes during proof testing.

a. 2
b. 3
c. 4
d. 5

_____ 8. Under ASME guidelines, the access gate on a platform must be equipped with a _____.

a. steel lock
b. friction brake
c. slip-resistant handle
d. positive-acting device to keep it from opening accidentally

_____ 9. The total weight of the loaded personnel platform must *not* exceed _____ percent of the rated capacity for all crane configurations.

a. 50
b. 60
c. 70
d. 80

_____ 10. According to OSHA, a crane shall be equipped with a boom angle indicator, a device to indicate boom extended length, and a(n) _____.

a. proximity warning device
b. control lockout system
c. anti-two-block device
d. drum sensor

_____ 11. Fall protection is covered under _____.

a. *OSHA 1910.180*
b. *OSHA 1926.104*
c. *OSHA 1926.106*
d. *OSHA 1926.550*

_____ 12. If rotation-resistant rope is used, the slings must be capable of supporting, without failure, at least _____ times the maximum intended load.

a. 4
b. 5
c. 6
d. 10

_____ 13. When hoisting personnel near power lines, _____ do(es) *not* negate the need for power line shutdown.

a. the experience of personnel in the platform
b. operator experience
c. operator aids
d. synthetic slings

_____ 14. According to ASME, periodic inspections of platforms are to be conducted at least once every _____ months.

a. 3
b. 6
c. 9
d. 12

_____ 15. Even if power lines have been shut down, _____ must be installed at the operator's station and on the outside of the crane.

a. signs warning of electrocution or bodily injury
b. insulating material
c. rubber gloves
d. electric rods

_____ 16. Hooks on overhaul ball assemblies, lower load blocks, or other attachment assemblies must be _____.

a. open
b. closed and locked
c. closed
d. permanently mounted

_____ 17. Wire rope, shackles, rings, master links, and other rigging hardware must be capable of supporting, without failure, at least _____ times the maximum intended load applied or transmitted to the component.

a. 3
b. 4
c. 5
d. 6

_____ 18. If travel is done with a rubber-tired carrier, the chart capacity for lifts on rubber must be used for application of the _____ percent reduction of rated capacity.

a. 30
b. 40
c. 50
d. 60

_____ 19. If deficiencies are found and corrected on a platform after a proof test has been conducted, _____.

a. the platform must be taken out of service until the manufacturer is contacted for safety precautions
b. hoisting operations can begin
c. another proof test needs to be conducted
d. only one person may go up in the platform

_____ 20. Under ASME guidelines, the inspection paperwork for the last inspection performed on a platform is to be kept _____.

a. on the job site somewhere
b. with the platform
c. in the crane cab
d. with the manufacturer

_____ 21. A multi-leg wire rope sling is known as a(n) _____.

a. bridle sling
b. anchorage sling
c. multiple sling
d. master link

_____ 22. A(n) _____ is a fastening or coupling device.

a. bridle
b. shackle
c. anchorage
d. lanyard

_____ 23. A full-body harness lanyard must have a maximum length that allows a fall of *no* greater than _____ feet.

a. 3
b. 6
c. 9
d. 11

_____ 24. A live boom is a boom in which _____.

a. the planetary pawl has been rendered inoperative
b. the hoisting action can be deactivated before damage occurs in the event of contact between the load block and the boom tip
c. lowering is regulated by a system or device on the power train, in addition to the load hoist brake
d. lowering is controlled by a brake without aid from other devices that slow the lowering speed

_____ 25. All full-body harness and lanyard hardware, *except* rivets, must be capable of withstanding a tensile loading of _____ pounds without cracking, breaking, or taking a permanent deformation.

a. 1,000
b. 2,400
c. 4,000
d. 5,400

_____ 26. If a personnel platform includes a perforated steel enclosure, the perforations must be *no* greater than _____ inch.

a. ¼
b. ½
c. ¾
d. 1

_____ 27. The toe board on a personnel lifting platform must be installed _____ of the platform.

a. on the gate
b. at the front and rear
c. on all sides except the gate side
d. around the entire perimeter

_____ 28. The top rail, toe board, and lower barrier on a personnel lifting platform must be capable of withstanding a load of _____ pounds applied at any point in an outward or downward direction.

a. 100
b. 300
c. 1,000
d. 2,000

_____ 29. The suspension system on a personnel lifting platform must have _____.

a. rope sling systems with mechanically spliced flemish eyes
b. synthetic webbing
c. synthetic fiber rope slings
d. wire rope clips and wedge sockets and knots

_____ 30. A boom angle indicator is required on all cranes with _____ booms.

a. lattice
b. telescoping
c. variable angle
d. live

Craft: Mobile Crane Operations

Module Number: 21305-05

Module Title: Hoisting Personnel

contren® Learning Series

TRAINEE NAME: ______________________________

TRAINEE SOCIAL SECURITY NUMBER: ______________________________

CLASS: ______________________________

TRAINING PROGRAM SPONSOR: ______________________________

____

INSTRUCTOR: ______________________________

Rating Levels: (1) Passed: performed task (2) Failed: did not perform task
Also, list the date the testing for each task was completed.

Recognition: When testing for the NCCER Standardized Craft Training Program, be sure to record Performance testing results on Craft Training Report Form 200, and submit the results to the Training Program Sponsor.

Objective	TASK	RATING
3	1. Visually inspect the platform, suspension system, and attachment points.	
4	2. Perform a test lift following appropriate safety procedures and regulations, including the use of fall protection.	
2, 4	3. Perform operations for hoisting personnel.	

notes

ANSWER KEY

Answer	Section Reference
1. d	3.4.0
2. a	5.1.0
3. d	7.1.0
4. a	2.0.0
5. a	4.0.0
6. d	3.4.0
7. d	7.1.0
8. d	4.0.0
9. a	5.1.0
10. c	5.2.0
11. b	2.0.0
12. d	3.4.0
13. c	8.0.0
14. d	6.1.0
15. a	8.0.0
16. b	3.4.0
17. c	3.4.0
18. c	5.4.0
19. c	7.1.0
20. b	6.1.0
21. a	3.4.0
22. b	3.4.0
23. b	2.0.0
24. d	5.1.0
25. c	2.0.0
26. b	3.2.0
27. d	4.0.0
28. b	4.0.0
29. a	4.0.0
30. c	5.2.0

notes

Lattice Boom Assembly and Disassembly

21306-05

Trainee Name: ____________________

Social Security Number: ____________________ **Date:** ____________

_____ 1. For large lattice boom cranes, the boom, counterweights, and _____ are removed for transportation.

a. mast
b. gantry
c. crawler treads
d. cab

_____ 2. The crane assembly area should be large enough to accommodate the crane and _____.

a. rigging equipment
b. the jib
c. the full arc of the boom
d. unloading trailers

_____ 3. When the crane swings, clearance between the counterweight and the nearest obstacle must be at least _____.

a. 1 foot
b. 2 feet
c. 3 feet
d. 4 feet

_____ 4. Moving _____ can cause mast failure if the weight exceeds the manufacturer's specifications.

a. counterweights
b. a boom section
c. crawler treads
d. the jib

_____ 5. Failure to remove additional counterweights before lifting operations will reduce the _____.

a. lifting capacity
b. swing radius
c. center of gravity
d. backward stability of the crane

_____ 6. *Never* add or remove counterweights from a carrier-mounted crane without first _____.

a. extending all outrigger beams and jacks
b. having a spotter to guide the load
c. engaging the emergency brakes
d. chocking all the tires

_____ 7. Nuts, bolts, and pins must be _____.

a. of the correct size and quality specific for the crane
b. of any kind the specific manufacturer makes
c. interchangeable
d. welded together

_____ 8. When assembling a lattice boom, you should _____.

a. find the shortest length that will accomplish the job
b. assemble the boom and all lifting attachments to make sure the crane is prepared for any lift
c. assemble the heel section last
d. assemble the boom before you get to the job site to speed up the lift process

_____ 9. Lattice boom sections can be handled using _____.

a. nylon slings
b. chainfalls
c. come-alongs
d. beam clamps

_____ 10. When using wire or chain slings, stay clear of _____ to prevent damage.

a. the cab
b. lacing
c. the crawlers
d. hydraulic cylinders

_____ 11. When assembling a lattice boom according to the procedure discussed in this module, you should *not* _____.

a. mix and match boom sections from different manufacturers
b. set the swing lock
c. extend the gantry
d. rig the main hoist line

_____ 12. Prior to assembly, rotate the upper works of the crane to face _____.

a. the front
b. the rear
c. over the side
d. the direction of maximum stability

_____ 13. The operator must make sure enough _____ are installed to lift the boom off the ground.

a. counterweights
b. lift pumps
c. undercrawler rods
d. pendants

_____ 14. The boom must *never* be elevated until _____ been erected.

a. the gantry or mast has
b. slings have
c. a hook has
d. a hoist line has

_____ 15. What section of the lattice boom gets installed first?

a. The base.
b. The intermediate section.
c. The swing-away, if it is being used.
d. The jib, if it is being used.

_____ 16. When the boom is assembled, check that the _____ is/are of the proper length for the boom length.

a. gantry
b. pendants
c. outriggers
d. pendant spreader bars

_____ 17. When assembling lattice boom sections, the pendants should be _____.

a. in front of the sections being connected
b. behind the boom sections being connected
c. on the ground directly under the boom
d. attached to the mast until all boom sections have been connected

_____ 18. Always install pendants _____.

a. clockwise
b. top-to-bottom
c. bottom-to-top
d. in the sequence noted on the rigging drawing

_____ 19. If the pendants are slightly long on a machine with a short boom, the _____.

a. weight of the mast will pull an unloaded boom into the back stops when the minimum radius is approached
b. boom will bend over forward, possibly overturning the crane
c. hoist rope will be damaged
d. mast will be damaged due to an overload condition

_____ 20. A chord is a(n) _____.

a. bar with fixed end connections
b. boom gantry
c. adjustable bar used to keep pendants from scuffing
d. loadbearing member of a lattice boom

_____ 21. In order to confirm proper assembly, once the boom is in the air, the _____ needs to be checked.

a. proximity-warning device
b. cold-start function
c. boom hoist limiting device
d. rotational lacing device

_____ 22. A rigid structural member projecting from a vertical support so that the upper part is in tension and the lower part is in compression is a _____.

a. cantilever
b. mast arm
c. boom hoist
d. spreader

_____ 23. Check that the pendants are the proper length _____.

a. after the boom is assembled
b. before the boom is assembled
c. after the jib is attached
d. after the boom base is attached

_____ 24. The boom should be raised by _____.

a. paying out the hoist line
b. raising the boom angle indicator
c. lifting the load block
d. engaging the boom hoist pawl

_____ 25. The load block should *not* be lifted when raising the boom _____.

a. to increase the speed of the lift
b. to mark a secure area in case of failure
c. so the load block can be reeved afterwards
d. to gain extra stability

_____ 26. When adjusting the pendants, the boom should be lowered onto the blocking until the _____.

a. hoist line is slack
b. total boom weight is supported by the blocking
c. pendants are slack
d. boom tip just rests on the blocking

_____ 27. The final steps in assembling the boom are _____.

a. reeve the block, raise the boom, and rig the main hoist line
b. move pendants to end of boom, reeve the block, and raise the boom
c. reeve the block, move pendants to the end of the boom, and raise the boom
d. rig the main hoist line, move pendants to the end of the boom, and raise the boom

_____ 28. If the pendants are too short on a machine with a longer boom, _____.

a. the boom, mast, pendants, and boom hoist systems will be overloaded when handling near-capacity loads
b. the boom will bend backward, possibly overturning the crane
c. the anti-two-block device will be damaged
d. the hoist rope will be damaged

_____ 29. The boom should be assembled _____.

a. in the order specified by the manufacturer
b. with the short sections close to the boom foot
c. with the short sections first
d. according to supervisor's instructions

_____ 30. A _____ is a strut member of a boom structure that is used to add structural support to a boom.

a. lacing
b. chord
c. weld
d. pendant

_____ 31. When disassembling the crane, drive out the lower pins _____the upper pins.

a. after driving out
b. before driving out
c. before inserting
d. after inserting

_____ 32. The jib should be assembled on _____.

a. the ground
b. blocking
c. a flatbed
d. a hill

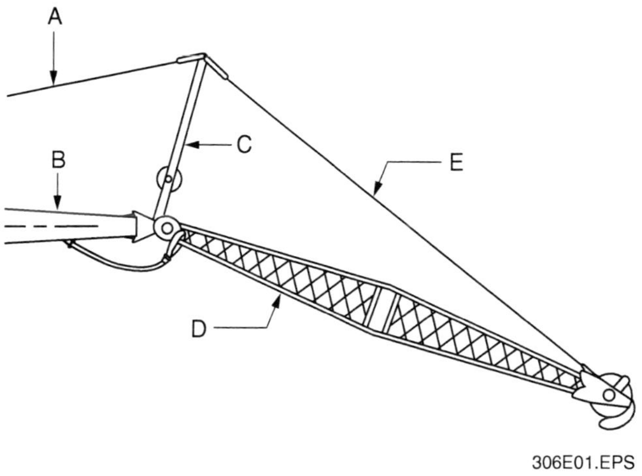

Figure 1

_____ 33. The boom section marked C in *Figure 1* is the _____.

a. boom tip
b. jib strut
c. jib backstay pendant
d. jib adapter

_____ 34. The boom section marked A in *Figure 1* is the _____.

a. boom tip
b. jib strut
c. jib backstay pendant
d. jib adapter

_____ 35. Before booming down for disassembly, you should _____.

a. adjust the pendant
b. drive out the lower pins
c. thoroughly inspect the boom hoist cable
d. lower the block to the ground

notes

Craft: Mobile Crane Operations

Module Number: 21306-05

Module Title: Lattice Boom Assembly and Disassembly

contren® Learning Series

TRAINEE NAME: ______________________________

TRAINEE SOCIAL SECURITY NUMBER: ______________________________

CLASS: ______________________________

TRAINING PROGRAM SPONSOR: ______________________________

____

INSTRUCTOR: ______________________________

Rating Levels: (1) Passed: performed task (2) Failed: did not perform task
Also, list the date the testing for each task was completed.

Recognition: When testing for the NCCER Standardized Craft Training Program, be sure to record Performance testing results on Craft Training Report Form 200, and submit the results to the Training Program Sponsor.

Objective	TASK	RATING
3	1. Determine the amount of counterweight required for lattice boom erection.	
2	2. Identify lattice boom components.	
4	3. Assemble and disassemble a short lattice boom.	
5	4. Assemble and disassemble a long lattice boom.	
6	5. Assemble a jib at the lattice boom top.	

notes

ANSWER KEY

Answer	Section Reference
1. c	1.0.0
2. d	2.0.0
3. b	2.0.0
4. a	3.0.0
5. d	3.0.0
6. a	3.0.0
7. a	4.0.0
8. a	4.0.0
9. a	4.0.0
10. b	4.0.0
11. a	5.1.0
12. d	5.1.0
13. a	5.1.0
14. a	5.1.0
15. a	5.1.0
16. b	5.1.0
17. b	5.1.0
18. d	5.1.0
19. a	5.1.0
20. d	4.0.0; Glossary
21. c	5.1.0
22. a	5.2.0; Glossary
23. a	5.1.0
24. d	5.1.0
25. d	5.2.0
26. c	5.2.0
27. b	5.2.0
28. a	5.2.0
29. a	5.2.0
30. a	4.0.0; Glossary
31. b	7.0.0
32. b	6.0.0
33. b	6.0.0
34. c	6.0.0
35. c	7.0.0

notes

Emergency Procedures

21307-05

Trainee Name: ______________________________

Social Security Number: ______________________ **Date:** ______________

_____ 1. When operating near minimum radius, the operator should boom down as the load is set down to compensate for _____ in the boom.

a. extension
b. overload
c. friction
d. elasticity

_____ 2. The ratio of electric intensity to a cross-sectional area is described as _____.

a. resistivity
b. elasticity
c. differential
d. fractional conductivity

_____ 3. If *no* fire has occurred after power line contact, it is all right to get off the crane _____.

a. after 10 minutes; the voltage will have reduced significantly so that it would pose no threat
b. after 20 minutes; the voltage will have been decreased to a level that would pose no threat
c. after electrical authorities de-energize the circuit and confirm that conditions are safe
d. one minute after the last spark was witnessed

_____ 4. If wire rope has been damaged by power line contact, it will _____.

a. be split
b. look like it was burnt by a torch
c. not move freely through the sheave and drum
d. be bright red

_____ 5. If you jump off a crane that has hit a power line, you should _____.

a. lean on the crane for balance
b. take large steps away from the crane
c. not contact the crane
d. stay in one spot

_____ 6. As part of safe mobile crane operations, you should verify the operator is qualified under _____ criteria.

a. *ASME B30.2*
b. *ASME B30.6*
c. *ASME B30.5*
d. *ASME B30.7*

_____ 7. To avoid backward collapse of a crane, the operator should _____.

a. pull continually on the hoist line after two-blocking
b. snub the hook block to the boom foot, then pull it up tight
c. carry the boom at the highest possible boom angle when traveling on an incline
d. boom down while setting the load down when operating near minimum radius

_____ 8. Operation of a(n) _____ can be affected by the different arrangement of multiple power lines, the movement of trucks, materials, or the crane itself.

a. proximity warning device
b. anti-two-block device
c. load angle indicator
d. drum rotation indicator

_____ 9. In a fire extinguisher description, the B stands for _____.

a. flammable liquids
b. chemical
c. caustic
d. dielectric

_____ 10. _____ need to be installed at the operator's station, and on the outside of the crane, warning that electrocution or bodily injury may occur unless appropriate clearance is maintained.

a. Durable signs
b. Rubber covers
c. Insulating foot pads
d. Grounding poles

_____ 11. The _____ must be held responsible for those operations under the operator's direct control.

a. operator
b. site supervisor
c. job contractor
d. signal person

_____ 12. To ensure safe lifting operations, the individual in charge of the job should demand that _____ inspection forms be provided with the crane.

a. quarterly
b. frequent and periodic
c. semi-annual
d. initial

_____ 13. The _____ is responsible for de-energizing power lines.

a. power company
b. job contractor
c. site supervisor
d. operator

_____ 14. If the crane is operating near the prohibited zone, _____ must be used for load control.

a. non-conductive tag lines
b. two signal people
c. a rope on a second drum on the crane
d. a second rope on the drum

_____ 15. When operating a crane within the avoidance zone, no one must be permitted to touch the crane or load unless the _____ indicates it is safe to do so.

a. signal person
b. tagline handler
c. operator
d. site supervisor

_____ 16. ASME standards require that a portable container for crane refueling be equipped with a(n) _____.

a. certificate of conformity
b. emission reduction system
c. flame arrester
d. secondary vent hole

_____ 17. The avoidance zone is _____.

a. an area in which crane operation is prohibited due to soil conditions
b. the quadrant the crane is not allowed to work in
c. the area near a power line that no part of the crane or load can enter
d. an unofficial zone in which limited crane operations can be performed if absolutely necessary

_____ 18. With the boom at a high angle, elasticity in the boom and boom hoist systems could cause _____.

a. the boom to move back against the boom stops when a load is released
b. snubbing of the hook block to the boom foot
c. boom loading to change from compression to bending
d. two-blocking

_____ 19. An arrester is a device that _____.

a. reduces the elasticity of the boom
b. prevents snubbing of the hook block to the boom foot
c. stops flashback ignition and may have a release mechanism that vents pressure above a certain psi
d. warns crane operators of power-line hazards

_____ 20. For voltages over 50 kV, the minimum required high-voltage power line clearance increases 5 feet for every _____ kV.

a. 50
b. 100
c. 150
d. 200

_____ 21. Boom guards, insulating links, and proximity warning devices are used to reduce the risk of _____.

a. backward collapse of a crane
b. crushing hazards
c. electrocution
d. injury from a falling load

_____ 22. The effective range of the fire extinguisher located in a mobile crane cab is _____ feet.

a. 1 to 7
b. 5 to 10
c. 5 to 20
d. 10 to 30

_____ 23. The crane or load may be grounded to line neutral by the utility before a crane operation that takes place with _____.

a. power lines energized and the crane operating less than the fully extended boom length away
b. power lines de-energized and grounded
c. power lines energized and the crane within the avoidance zone
d. the crane in transit with the boom lowered

_____ 24. If power line contact does occur, the operator of the crane should _____.

a. step down from the cab
b. remain in the cab if there is no danger of fire or explosion
c. jump from the cab and take steps that are as large as possible
d. jump from the cab and run away as fast as possible

_____ 25. A boom and load must *not* be operated _____.

a. near minimum radius
b. in the prohibited zone
c. within the avoidance zone
d. with the crane/load grounded to line neutral

ANSWER KEY

Answer	Section Reference
1. d	5.2.0
2. a	4.5.0; Glossary
3. c	4.5.0
4. b	4.5.0
5. c	4.5.0
6. c	2.0.0
7. d	5.2.0
8. a	4.4.0
9. a	3.0.0
10. a	4.1.0
11. a	2.0.0
12. b	2.0.0
13. a	4.0.0
14. a	4.2.0
15. a	4.3.0
16. c	3.0.0
17. d	4.0.0
18. a	5.2.0
19. c	3.0.0
20. c	4.0.0
21. c	4.1.0
22. c	3.0.0
23. c	4.3.0
24. b	4.5.0
25. b	4.0.0

notes

Transporting Requirements

21308-05

Trainee Name: ______________________________

Social Security Number: ______________________ **Date:** __________

_____ 1. When loading crawler cranes on a flatbed, the ramp must be _____ enough to make the ramp angle _____.

a. long; high
b. long; low
c. short; high
d. short; low

_____ 2. When lifting lattice boom sections onto flatbeds, slings should be _____.

a. attached at the location of the pin inserts
b. attached at opposite corners of the lattice parts
c. wrapped around the lattice once and then attached to the pin inserts
d. attached to the fittings at the ends of the main chords

_____ 3. Boom sections should be secured to the trailer with _____.

a. nylon straps
b. metal chains
c. cable
d. steel braces

_____ 4. Internal bridge weight limits apply to the weight allowed on _____.

a. each axle
b. the Interstate Highway System
c. tires
d. overpasses

_____ 5. According to Federal DOT regulations, the maximum height that a truck/trailer combination and its load can be is _____.

a. 12 feet, 6 inches
b. 13 feet, 6 inches
c. 14 feet, 6 inches
d. 15 feet, 6 inches

_____ 6. When traveling a truck crane, the position of the boom should be _____.

a. in line over the front with the forward direction of the crane
b. facing towards the forward direction of the crane at a 10-degree angle
c. in line with the opposite direction the crane is traveling
d. extended halfway to its maximum length in the forward direction of travel

_____ 7. A _____ should be used to release a binder.

a. steel bar
b. cable
c. handle extension
d. cheater pipe

_____ 8. Commercial truck-mounted cranes must meet the requirements of the _____.

a. American Society of Mechanical Engineers
b. Occupational Safety and Health Administration
c. Department of Motor Vehicles
d. Department of Transportation

_____ 9. With regard to motor carrier regulations, which of the following is a true statement?

a. The operator should always obey local regulations, rather than DOT regulations.
b. The operator should always obey DOT regulations and ignore local regulations.
c. If the DOT imposes a higher standard than the local ordinance, then the DOT regulation must be complied with.
d. The operator does not have to take a physical exam as part of driving eligibility requirements.

_____ 10. When loading crawler cranes onto a flatbed, the _____ must be long enough to make the ramp angle low.

a. boom
b. crawlers
c. ramp
d. trailer

_____ 11. The procedures for loading and transporting a mobile crane depend on _____.

a. the number of trucks available
b. the space available for loading
c. company policy
d. the size and weight of the crane

_____ 12. Boom extensions stacked on a trailer should be _____.

a. of equal length
b. placed at 45-degree angles and rigged together
c. placed at 10-degree angles and blocked to prevent friction
d. placed with the larger section on top

_____ 13. Ratchet-type binders multiply stored energy by a factor of _____.

a. 30
b. 35
c. 40
d. 50

______ 14. The maximum load allowed on a tandem axle is ______ pounds.

a. 20,000
b. 34,000
c. 44,000
d. 80,000

______ 15. If the total load for a truck and trailer combination is more than 80,000 pounds you must ______.

a. reload the trailer to reduce the weight
b. obtain a permit
c. reroute to avoid bridges
d. travel only on back roads

notes

Craft: Mobile Crane Operations

Module Number: 21308-05

Module Title: Transporting Requirements

contren®
Learning Series

TRAINEE NAME: ______________________________

TRAINEE SOCIAL SECURITY NUMBER: ______________________________

CLASS: ______________________________

TRAINING PROGRAM SPONSOR: ______________________________

____

INSTRUCTOR: ______________________________

Rating Levels: (1) Passed: performed task (2) Failed: did not perform task
Also, list the date the testing for each task was completed.

Recognition: When testing for the NCCER Standardized Craft Training Program, be sure to record Performance testing results on Craft Training Report Form 200, and submit the results to the Training Program Sponsor.

Objective	TASK	RATING
4	1. Properly load a crane and its components onto a flatbed trailer.	
5	2. Properly secure a load on a flatbed trailer.	

notes

ANSWER KEY

Answer	Section Reference
1. b	2.1.0
2. d	2.1.0
3. a	2.1.0
4. b	4.0.0
5. b	4.0.0
6. a	3.0.0
7. a	2.2.0
8. d	3.0.0
9. c	4.0.0
10. c	2.1.0
11. d	2.0.0
12. a	2.1.0
13. d	2.2.0
14. b	4.0.0
15. b	4.0.0

notes